Progress in Drug Research

Founded by Ernst Jucker

Progress in Drug Research

Peptide Transport and Delivery into the Central Nervous System

Vol. 61

Edited by
Laszlo Prokai and
Katalin Prokai-Tatrai

Springer Basel AG

Editors:

Prof. Laszlo Prokai
Department of Medicinal Chemistry
College of Pharmacy and
The McKnight Brain Institute
University of Florida
1600 S.W. Archer Road
Gainesville, FL 32610-0497
USA

Prof. Katalin Prokai-Tatrai
Department of Pharmacology and
Therapeutics
College of Medicine
University of Florida
1600 S.W. Archer Road
Gainesville, FL 32610-0267
USA

Visit our PDR homepage: http://www.birkhauser.ch/books/biosc/pdr

ISBN 978-3-0348-9420-3   ISBN 978-3-0348-8049-7 (eBook)
DOI 10.1007/978-3-0348-8049-7

9 8 7 6 5 4 3 2 1

Contents

Foreword by the Editors

Hypo- or hypersecretion, alteration in storage, release, catabolism, and post-translational processing of neuropeptides are associated with the etiology of many diseases affecting the central nervous system (CNS). Various peptides native to the brain and the spinal cord, as well as synthetic peptides, peptide analogues and peptidomimetics developed as their agonists or antagonists could be useful in the treatment of these CNS maladies. However, peptides face a formidable obstacle in reaching the intended site of action due to the existence of the blood-brain barrier (BBB) – a vital element in the regulation of the internal environment of the brain and the spinal cord.

This theme-based volume of the *Progress in Drug Research* series entitled "Peptide Transport and Delivery into the Central Nervous System" consists of eight reviews by internationally known experts covering the most important aspects pertaining to the emerging field of peptide neuropharmacotherapy. First, Fleur Strand provides an overview about neuropeptides, highlighting their general characteristics necessary for understanding their role in the regulation of physiological systems and their neuropharmaceutical potential. David Begley and Milton Brightman cover the morphology and properties of the BBB in terms of solute transport into the CNS and potential mechanisms to cross the BBB. Then, Abba Kastin and Weihong Pan review peptide transport across the BBB and highlight structural, biochemical and physiological features that govern the bidirectional movement of these biomolecules across the endothelial cells of the cerebral microcirculation.

Although a small portion of peripherally delivered peptide may reach the brain parenchyma, the amount is often insufficient to exert CNS effects. Therefore, methods to deliver these biomolecules into the brain and spinal cord are necessary. Delivering peptides into the CNS has been a challenging task and various approaches have been developed. The five subsequent reviews of the volume discuss the rationale, present state-of-the-art and future trends of various strategies applicable to peptides in order to outwit the BBB as the major obstacle of peptide pharmacotherapy of the CNS.

Invasive methods are covered first. Richard Grondin et al. detail direct intracranial delivery of proteins and peptides with special emphasis on potential treatment of neurodegenerative diseases. David Fortin focuses on approaches that involve transient opening of the tight junctions of the endothelial cells, which is clinically used in brain-tumor therapy, or permeabilizing the endothelial cells separating the systemic circulation from the interstitial fluid of the CNS.

Noninvasive approaches for the potential targeting of peptides into the CNS are covered by three reviews. First, we discuss prodrug strategies that aim at bioreversibly altering the properties of the peptide to enhance BBB transport. Subsequently, a comprehensive review by Suresh Vyas covers physiologically based strategies that utilize biological carriers to gain access to the CNS. Finally, Jamal Temsamani and Jean-Michel Scherrmann highlight specific peptides that may actually be utilized as vectors to ferry therapeutic molecules manifesting poor BBB-transportability into the brain and spinal cord.

All the eight reviews in the volume contain extensive bibliographies enabling the interested reader to use this volume as an encyclopedic source of information on intriguing issues of turning neuropeptides to potential drugs of the (hopefully near) future for treating CNS-maladies.

Our very special thanks go to the contributing authors and to our publisher whose concerted efforts made this volume possible. We hope it will give the reader as much enjoyment as we had putting it together.

Gainesville (Florida), May 2003
Laszlo Prokai
Katalin Prokai-Tatrai

Progress in Drug Research, Vol. 61
(L. Prokai and K. Prokai-Tatrai, Eds.)
©2003 Birkhäuser Verlag, Basel (Switzerland)

Neuropeptides: general characteristics and neuropharmaceutical potential in treating CNS disorders

By Fleur L. Strand

New York University
340 East 64th Street
New York, NY 10021, USA
<fleur.strand@nyu.edu>

Fleur L. Strand

received her Ph.D. in Biology from New York University (1952). Her post-doctoral research on the effects of stress on the neuromuscular system led to her discovery that ACTH and its peptide fragments have direct effects on peripheral nerve regeneration and the development of motor pathways. Her later research demonstrated the neuroprotective and neurotrophic action of melano-cortins on both central and peripheral motor systems. She is the author of five textbooks, co-editor of more than seven Annals of the New York Academy of Sciences and has published more than 80 articles in refereed journals. Her publications also include invited chapters in several books. Her physiology textbook received the American Medical Writers' award in 1982. Her most recent book is Neuropeptides: Regulators of Physiological Processes *(MIT Press, 1999). Dr. Strand has chaired the Winter Neuropeptide Conference in Breckenridge, CO, for many years and is on the advisory board of the Summer Neuropeptide Conference. She is a founding member and Secretary of the International Neuropeptide Society and has been an invited speaker at many international and national meetings. She has also acted as consultant for several pharmaceutical companies. Dr. Strand was President of the New York Academy of Sciences (1976). She chaired the Mayor's Award for Excellence in Science and Technology: Biology and Medicine section (1996-2001). She was selected as Outstanding Woman Scientist by the New York chapter of AWIS in 1987. She serves on the New York State Spinal Cord Injury Board and on the Editorial Boards of the journals* Peptides *and* Regulatory Peptides. *Dr. Strand is presently Carroll and Milton Petrie Professor Emerita of Biology and Neural Science at New York University.*

Contents

Key words

Neuropeptide characteristics, distribution in CNS, energy homeostasis, neuropeptides and appetite regulation, CNS disorders, inflammation, neuroregeneration, neuroprotection, translational and combinational research.

Glossary of abbreviations

ACh, acetylcholine; ACTH, adrenocorticotropic hormone; AD, Alzheimer's disease; AgRP, agouti-related protein; Arg, arginine; CGRP, calcitonin-gene related peptide; CCK, cholecystokinin; CNS, central nervous system; CRH, corticotropin-releasing hormone; DRG, dorsal root ganglia; ER, endoplasmic reticulum; GABA, γ-aminobutyric acid; GH, growth hormone; GHRH, growth hormone releasing hormone; HPA, hypothalamic-anterior pituitary-adrenal axis; IGF-I, insulin-like growth factor; Lys, lysine; MSH, melanocyte-stimulating hormone; NMDA, N-methyl-D-aspartate; OT, oxytocin; PC, prohormone convertase; POMC, proopiomelanocortin; PPP, pancreatic polypeptide; PRL, prolactin; RER, rough endoplasmic reticulum; ST, somatostatin; TRH, thyrotropin-releasing hormone, VIP, vasoactive intestinal peptide; VP, vasopressin.

1　General characteristics of neuropeptides

1.1　The neuropeptide concept

Earlier concepts of the neuroendocrine system integrated the regulatory functions of the nervous and endocrine systems, a concept that was developed from the understanding of hypothalamic control over the secretions of the anterior pituitary gland and pituitary control of its target endocrine organs. Demonstration of feedback controls, chiefly negative, furthered the argument for an intimate relationship between the nervous and endocrine systems. However, the role of the hypothalamic and pituitary hormones, all of which are peptides, is not restricted to their classical endocrine actions but extends to profound effects on the CNS. Experiments showing that adrenocorticotropic hormone (ACTH) affects complex behavioral processes such as learning and memory in adrenalectomized rats, demonstrated clearly that this peptide was acting directly on CNS neurons, bypassing its endocrine target, the adrenal cortex. This concept of a direct action was extended to learning and memory in humans, as well as to the peripheral neuromuscular system through the use of synthetic fragments of ACTH devoid of adrenocortical-stimulating activity [1–4] (see sections 5.1.2 and 5.4).

In addition, many of the hypothalamic and pituitary hormones are produced by other brain areas, the gastrointestinal tract and by lymphocytes. The remarkable diversity of production sites of mammalian neuropeptides is shown in Table 1 [5].

The idea that peptides produced in the brain and gut have direct effects on neurons and on many vital physiological processes took many years to gain the acceptance that it presently receives.

Neuropeptides affect non-neural tissues and organs as well as neurons and bear an important responsibility for the integration of brain functions and the systems of the body. They may act as neurohormones or neurotransmitters and frequently are co-localized with the classical neurotransmitters such as acetylcholine and the monoamines. They are involved in the regulation of reproduction, growth, water and salt metabolism, temperature control, food and water intake, cardiovascular, gastrointestinal and respiratory control, behavior, memory and affective states. Neuropeptides are involved in many autonomic responses and they potently modulate nerve development and regeneration. Consequently high hopes have been raised for the clinical application of neuropeptide therapy for disorders or trauma involving physiological systems but these trials have in general proved disappointing. Multiple neuropeptides may regulate a specific physiological process such as food intake, sexual behavior or metabolism. Neuropeptides may only be released in significant amounts during development, regeneration or under pathological conditions. Knockout mice, in which a specific gene for a neuropeptide is deleted, often show little change of phenotype, indicating that redundant systems probably exist. There is an enormous potential for neuropeptides in the treatment of CNS disorders but the challenge is to find the combination of neuropeptides necessary to add or eliminate in order to restore the physiological equilibrium and offset CNS pathology. This requires knowledge of the general characteristics of neuropeptides as well as a detailed understanding of the specific properties of each neuropeptide being investigated. Another important consideration is the responsiveness of tissues to neuropeptide action, a characteristic that varies considerably during ontogeny, development and maturation [6]. Genes that are expressed early in development may be turned off as the organism matures, but re-expressed following a change in metabolism or a severe injury [7].

1.2 Evolution of neuropeptides

Neuropeptides such as insulin, ACTH, cholecystokinin (CCK), gonadotropin, β-endorphin and many others, are found in unicellular animals and plants, invertebrates and in vertebrates ranging from lampreys to humans. The bio-

Table 1.
Mammalian neuropeptides classified according to principal source (other sources in right-hand column)

Hypothalamic peptides	
Agouti-related protein	
β-endorphin	pituitary
corticotropin releasing hormone	pituitary, median eminence, brain, spinal cord
galanin	locus ceruleus
gonadotropin releasing hormone	placenta, gonads?
growth hormone releasing hormone	placenta
melanin concentrating hormone	
orexin/hypocretin	
oxytocin*	posterior pituitary, brain, spinal cord
pituitary adenylate cyclase activating polypeptide	thalamus, hippocampus, spinal cord
proopiomelanocortin	pituitary
somatostatin	brain, spinal cord, gut, salivary glands, excretory system
thyrotropin releasing hormone	brain, spinal cord, gut, pancreas
Tyr-MIF-1	cerebral cortex
vasopressin*	posterior pituitary, brain, spinal cord, peripheral nerves

*produced in the hypothalamus but stored in the neurohypophysis and called neurohypophyseal peptides

Anterior pituitary peptides	
adrenocorticotropin	median eminence, brain, spinal cord, placenta, gut, lymphocytes
β-endorphin	median eminence, placenta, gut
follicle stimulating hormone	median eminence, placenta
growth hormone	median eminence, placenta
luteinizing hormone	median eminence, placenta
melanocyte stimulating hormone	median eminence, brain, spinal cord, peripheral nerves
prolactin	median eminence, brain, spinal cord, placenta
thyroid stimulating hormone	median eminence, placenta

Brain and spinal cord peptides	
δ sleep-inducing peptide	
dynorphin	
endorphins	pituitary
Met- and Leu-enkephalin	myenteric neurons, peripheral nerves
neuropeptide glutamine (E)isoleucine-(I)	
neurotensin	cardiovascular system, gastrointestinal tract
nociceptin/orphanin	brain stem, trigeminal ganglion, CNS?
secretoneurin	

Major gut and pancreatic peptides	
ghrelin	stomach, intestine, placenta, pituitary, hypothalamus?
calcitonin gene related peptide – GI tract	brain, spinal cord, neuromuscular junctions, thyroid
cholecystokinin – intestine	brain, pituitary, adrenal medulla, peripheral nerves
galanin – intestine	brain, spinal cord, almost all peripheral systems and organs
gastrin – gastric antrum	pituitary, spermatozoa

Table 1. (continued)

gastric inhibitory peptide – duodenum	
gastrin releasing peptide* – GI tract	spinal sensory ganglia, spinal cord, brain
glucagon – intestine	brain, median eminence
insulin – pancreas	neonatal brain
motilin – small intestine	brain, peripheral nerves?
neuropeptide Y- myenteric neurons	brain, peripheral nerves, adrenal medulla
peptide histidine isoleucine – intestine	cerebral cortex, hypothalamus
neuropeptide YY – myenteric neurons	brain stem
neurotensin – intestine	median eminence, hypothalamus, pituitary, entire CNS
opioids – myenteric neurons	brain, spinal cord, peripheral nerves
pancreatic polypeptide – pancreas	
secretin – intestine	brain, spinal cord, pituitary, pineal
somatostatin – gastric cells	hypothalamus, brain spinal cord, thyroid, excretory system
substance P – intestine	brain, spinal cord, peripheral nerves
vasoactive intestinal polypeptide – myenteric neurons	cerebral cortex, hypothalamus, peripheral nerves
* bombesin-like peptide	
Placental peptides	
chorionic gonadotropin	
corticotropin releasing hormone	pituitary
follicle stimulating hormone	ovary
growth hormone	pituitary
growth hormone releasing hormone	pituitary
inhibin	gonads
luteinizing hormone	ovary
prolactin	pituitary
proopiomelanocortin	hypothalamus, pituitary
somatostatin	brain, spinal cord, gut, salivary glands, excretory system
thyrotropin-releasing hormone	brain, spinal cord, gut, pancreas
Cardiac peptide	
atrial natriuretic hormone	hypothalamus
Blood peptide	
angiotensin II	hypothalamus, pons
Thyroid peptides	
calcitonin	brain
calcitonin gene related peptide	brain, spinal cord, neuromuscular junctions
somatostatin	hypothalamus, brain, spinal cord, excretory system
Parathyroid peptides	
parathyroid hormone	
parathyroid related protein	vascular smooth muscle, milk, tumors

Taken from [5], slightly modified by author.

logically active part of the structure of the molecule has been highly conserved throughout evolution, for example the mammalian insulins are very highly conserved with only one or two sequence differences between porcine, bovine and human insulin. This suggests that almost all of the insulin structure is essential for its functions, which include not only binding to its receptor but also regulation of the stereospecific processing of the biosynthetic precursor of the insulin molecule and the storage of insulin granules [8]. In contrast, there is little homology between rat and porcine relaxin, indicating considerable divergence between these two species. However, conservation of structure often is accompanied by a change in function of the peptide and its receptors. Prolactin (PRL), one of the oldest known neuropeptides, has changed its function radically throughout evolution. In all species, from fish to mammals, PRL is concerned with water and salt metabolism; in birds it is vital for the production of crop milk; in mammalian females it is responsible for lactation and in mammalian males, including humans, PRL is involved in spermatogenesis, testosterone synthesis and male libido. The common ancestry of many neuropeptides is reflected in the ability of neuropeptides derived from plants, amphibians, fish and lower mammals to activate physiological processes in humans: vasopressin from fish causes water retention in humans; morphine derived from a species of poppy produces the same effect in humans as the mammalian endorphins.

1.3 Gene duplication and neuropeptide families

Neuropeptides are formed from large pre-prohormones and prohormones, which are progressively split by specific proteolytic enzymes. The existence of families of neuropeptides indicates that they were generated through successive events of gene duplication within a common ancestral peptide, a process that permits a small error rate of mutation. Families such as the growth hormone (GH) family, insulins, enkephalins, the gut hormones and the tachykinins, have extensive amino acid sequence homology. Similarly, the necessary new receptors and processing enzymes probably evolved through gene duplication and show familial relationships as befits their neuropeptide ligands and substrates, respectively. This process has been described by Acher and Chauvet [9] as a molecular cascade of neuroendocrine control of organismal functions, resulting in the synchronous evolution of the precursor molecule, control of its passage into storage vesicles and secretion into the circulation, as well as the recognition of the circulating hormone of membrane receptors on its target cells.

The large precursor molecule, proopiomelanocortin (POMC), produced in the anterior pituitary gland, contains ACTH, β-endorphin and three similar amino acid sequences of α-, β-, and γ-melanocyte-stimulating hormone (MSH). These multiple copies of related molecules are best explained by the concept of gene duplication. The existence of distinct receptors for the various peptide fragments supports the concept of coordinated evolution. The feedback systems controlling the synthesis of neuropeptides also requires coordinated evolution.

Gene duplication in the neurohypophyseal family has been examined in all classes of vertebrates and each species has two neurohypophyseal nonapeptides: the oxytocin (OT)-like peptides involved in reproduction, milk ejection and uterine contractility; and the vasopressin (VP)-like peptides which have antidiuretic and pressor functions. The VP family has in common a basic amino acid at position 8, whereas the OT-related peptides have a neutral amino acid in this position. Probably, duplication of the gene followed by mutation of residue 8 in one of the genes, resulted in the dual evolution of the two lines of peptides and their separate receptors [10].

2 Biosynthesis, processing, secretion and inactivation of neuropeptides

2.1 Pre-prohormones and the signal peptide sequence

Neuropeptides ranging in size from 3 to 40 amino acid residues constitute a major group of intercellular agents for cell-to-cell communication, either as messenger hormones or as neurotransmitters and neuromodulators. The communication process begins within the neurosecretory cell with the expression of the promessenger gene. This is followed by the formation of the pre-prohormone, splitting off of the signal peptide to form the prohormone precursor, processing of the precursor into individual peptides, storage of the active form of the neuropeptide in vesicles, and finally, secretion of the neuropeptide. In turn, the response of the target cell depends on neuropeptide recognition of specific membrane receptors and the consequent transduction of the message through a second intracellular cascade to activate the effector. Interference, either positively or negatively, with any of these processes may be a route for therapies useful for clinical application.

The large pre-prohormone consists of the amino acid sequence of the neuropeptide, plus a signal sequence at its N-terminus that guides it through the ribosome and into the rough endoplasmic reticulum (RER). The signal peptide of about 15 to 30 amino acids is found on the precursors of hormones designed to be secreted. The pre-prohormones range in size from about 10 to 35 kDa.

The signal peptide that directs the pre-prohormone has a hydrophilic N-terminus and is followed by a peptide separated from the following peptide in the parent molecule by two basic amino acids, usually Lys-Arg. This is the site at which it is cleaved from the pre-prohormone by a signalase, an endopeptidase. The hydrophilic core of the signal sequence consists of about 10 hydrophobic amino acids a fixed distance from the cleavage site, which facilitates the insertion of the pre-prohormone into the lipid of the RER membrane. Blobel and Dobberstein [11] suggest that it is the interaction of the signal peptide and the membrane that recruits membrane receptors (transmembrane glycoproteins), causing them to associate through ionic bonds to form a tunnel, which is further stabilized by the binding of the 60S subunit of the ribosome to the membrane. The cleavage mechanism is highly conserved: endopeptidases from microsomal membranes of bacterial cells, amphibians and mammals can cleave signal sequences from the pre-proproteins from these diverse sources [10].

The large, biologically inactive precursor molecule, the prohormone, is synthesized on ribosomes and subsequently subjected to enzymatic proteolysis to yield active neuropeptides of varying size, characteristics and potencies. These precursors may contain one or more sequences for closely related peptides, as well as for unrelated peptides. The nascent precursors are closely associated with intracellular membranes, through which they are translocated, chemically processed, packaged and prepared for secretion. The excision of specific neuropeptides may vary according to the enzymes within specific tissues and the vesicles within which the enzymes are contained.

2.2 Prohormones and precursor processing

The peptide chain with the signal peptide removed is the biologically inert prohormone containing one or more copies of the final active neuropeptide. Folding of the protein occurs in the cistern of the endoplasmic reticulum

(ER), from which they are shuttled to the *cis*-face of the Golgi apparatus, then targeted to the *trans*-Golgi network, where packaging of the precursor into granules of the regulated secretory pathway occurs. Sections of the Golgi network filled with the prohormone bud off to form granules or vesicles. It is here that post-translational precursor processing occurs: the excision of peptide sequences by endoproteases to generate a diversity of biologically active peptides. The granules contain a series of co-packaged processing enzymes which sequentially attack the prohormone as the granules move from the cell body down to the nerve terminals. During their transit through the Golgi network, precursors may be subjected to many enzymatic modifications following initial endoproteolytic processing, such as glycosylation, phosphorylation, sulfation of tyrosyl residues or oligosaccharides, and hydroxylation of lysine residues [12].

Peptide bond hydrolysis of the prohormone usually occurs through a trypsin-like endopeptidase at precursor sites represented by pairs of basic amino acids arranged in doublets adjacent to the bioactive peptide sequences (Lys-Arg, Lys-Lys and Arg-Arg). There appears to be no consistent pattern for the position of the active fragments inside the prohormone. In proinsulin the active chains (A and B) are at the ends of the prohormone; in glucagons and calcitonin they are more centrally placed; proenkephalin contains several copies of met-enkephalin as well as one leu-enkephalin; and POMC contains several biologically active neuropeptides as well as several inactive fragments. Enzymes that cleave at paired basic residues have strict substrate specificity: if one amino acid is mutated, cleavage is totally blocked. However, cleavage at single arginine sites occurs in the biosynthesis of atriopeptin, somatostatin (ST), CCK, dynorphin, pancreatic polypeptide (PPP) and growth hormone releasing hormone (GHRH). Prosomatostatin contains both dibasic and monobasic sites. Cleavage is generated first by endopeptidases, followed by exoproteolysis by amino- and carboxypeptidases, and in some cases by special amidating enzymes to free the active neuropeptide.

The precursor processing enzymes have been identified as prohormone convertases (PCs) and form a family of seven mammalian proteases that bear a striking homology to a bacterial protease, subtilisin, indicating strong evolutionary conservation of these residues. Other modification of the prohormone may be post-transcriptional but pre-translational: calcitonin and calcitonin-gene related peptide (CGRP) seem to be derived from a single gene after alternate splicing at the mRNA level.

POMC was the first prohormone to be shown to be differentially processed in a tissue-specific manner to yield multiple active neuroepeptides. In the anterior pituitary lobe, cleavage of POMC yields chiefly a large N-terminal fragment, ACTH 1-39, and β-lipotropin. The prohormone convertose (PC) known as PC1/3 is responsible for this cleavage and also for a similar cleavage in the intermediate lobe. However, the intermediate lobe also contains a different PC, (PC2) which further hydrolyzes ACTH and β-lipotropin to the shorter neuropeptides: α-MSH (which is ACTH 1-13), γ-lipotropin and β-endorphin, respectively. Additional amino acid sequences (CLIP and JP) of unknown biological activity are also generated in the intermediate lobe. It is noteworthy that the human pituitary gland does not have an intermediate lobe. Consequently all processing of POMC in the human occurs within the anterior pituitary gland. The differential processing patterns of a prohormone may also depend on the relative amounts of the individual PCs within a neuroendocrine cell [13]. There are similar differences in the distribution of the PCs in the paraventricular nucleus and the supraoptic nucleus.

2.3   Exoproteolyis and neuropeptide segregation

Exoproteolysis follows the cleavage of the peptide precursor by endoproteolysis. The newly exposed C-terminal basic residues are removed by carboxypeptidase E which has a high specificity for basic residues. Peptides having an exposed C-terminal glycine residue often undergo conversion of this residue to an amide.

Most of the proteolytic processing of neuropeptides appears to occur within the secretory granules after they have been formed in the *trans*-Golgi network and the products of proteolytic cleavage are stored and secreted in equimolar amounts. Depending on the sequential processing of the prohormone, the specific neuropeptide content of the vesicles will vary; if the prohormone undergoes endoproteolysis before being packaged in a vesicle, the vesicle may contain only one of the active sequences of the prohormone. However, if endoproteolysis occurs within the vesicle, that vesicle may contain several different biologically active neuropeptides. This segregation of neuropeptides within different vesicles, or their coexistence within the same vesicle, is of considerable significance in determining the pattern of release of these regulatory agents after stimulation. This characteristic is another site to be targeted by drugs of clinical relevance.

2.4 Precursor genes

The organization of the POMC gene has been extensively studied and serves as a model for most neuropeptides. The mRNA coding for POMC and its translation products make up almost one-third of the total translation products in the bovine pituitary intermediate lobe [14, 15]. The basic structure of the POMC gene has been highly conserved, the most conserved regions being centered around each of the MSH units. The gene contains three exons and two introns, with the large 3′ exon containing the nucleotides coding for all the biologically active neuropeptides (ACTH and α-MSH, β-lipotropin, β-MSH, β-endorphin, and γ-MSH) and most of the N-terminal part of the precursor. Exon 1 includes the nucleotide sequence complementary to the 5′ sequence of POMC mRNA and is a non-coding exon. Exon 2 codes for the signal peptide and the first 18 amino acids of POMC. There is a repetition of the melanotropin core hepta-sequence Met-Glu-His-Phe-Arg-Tryp-Gly in ACTH, α-MSH, β-MSH and γ-MSH.

Neuropeptide gene expression is usually regulated by a complex series of negative and positive feedback controls. In the anterior pituitary lobe, the secretion of POMC is stimulated by corticotropin-releasing hormone (CRH) and VP, and inhibited by glucocorticoids from the adrenal cortex. In the intermediate lobe, the release of POMC peptides is inhibited by the neurotransmitters dopamine and γ-aminobutyric acid (GABA). These secretory changes are preceded by the appropriate increase or suppression of POMC mRNA activity in the pituitary. Hypothalamic POMC mRNA levels are also controlled by gonadal steroids, especially estrogen.

2.5 Excitation-secretion coupling and exocytosis

Neuroendocrine cells demonstrate a wide repertoire of electrical responses to secretagogues, mediated by voltage-gated sodium, calcium and potassium channels in the plasma membrane associated with a rise in intracellular calcium concentration $[Ca^{2+}]_i$. Some cell types generate one or two patterns whereas others, such as thyrotropin-releasing hormone (TRH)-stimulated lactotrophs in the anterior pituitary display almost all these electrical response patterns [16].

Exocytosis entails the formation of fusion pores between membrane-bound vesicles and the presynaptic membrane, and the subsequent release of the vesicular contents into the extracellular space. This process requires multiple Ca^{2+} binding proteins with different affinities and a Ca^{2+} sensor for the final fusion with the plasma membrane. Neuropeptides, which are contained within large (> 70 nm) dense-core vesicles, are subjected mainly to regulated exocytosis, which is triggered by a rise in cytosolic Ca^{2+} following voltage-gated Ca^{2+} influx through the plasma membrane. The complex characteristics of neuropeptide release from the large dense-core vesicles are described by Nicholls [17] and Burgoyne and Morgan [18].

2.6   Inactivation of neuropeptides

Very little degradation of neuropeptides occurs intracellularly, in contrast to their rapid inactivation in plasma. Some of the products of extracellular proteolytic cleavage are biologically inactive and many of the enzymatic modifications of the precursor may yield inactive metabolites. Extracellular inactivation through enzymatic degradation occurs rapidly in plasma and by other cells, including glia, due to the widespread presence of peptidases. Neuropeptides are rapidly inactivated in the gastrointestinal tract and are therefore ineffective when administered orally. Of clinical importance is that peripherally administered neuropeptides remain effective far longer than their biological half-life (about 12 to 15 minutes) due to their tenacious binding to cell membranes. This apparently protects them from proteolysis and permits them to perpetuate the cascade of receptor-mediated second messenger effects for several hours.

3   Distribution and localization of CNS neuropeptides

3.1   Distribution in the CNS

The richest sources of neuropeptides in the CNS are several hypothalamic nuclei, the preoptic area and the pituitary gland. However, many of the neurosecretory hypothalamic neurons have projection axons long enough to reach fairly distant areas of the CNS, so that these neuropeptides are found in high concentration in nerve terminals in many regions of the brain and spinal cord.

Other brain areas also contain neuropeptides. In addition to their concentration in hypothalamic nuclei and the pituitary gland, several neuropeptides such as ST, TRH, neurotensin (NT), substance P (SP), insulin, CCK, vasoactive intestinal peptide (VIP), and the enkephalins are distributed in the dorsal horn of the spinal cord, the dorsal vagal complex, the nucleus accumbens, the stria terminalis bed nuclei, the amygdala, and the periaqueductal gray. Unlike the extensive peptidergic pathways in the hypothalamic neurosecretory neurons, the pathways of neurons in extrahypothalamic brain areas are obscure and probably act in a paracrine manner on neighboring neurons. Many of these "brain" neuropeptides are widely distributed in various regions of the body, especially mucosal cells in the gastrointestinal tract, the pancreas, adrenal medulla, gonads, the placenta and peripheral nerves (Tab. 1).

3.2 Coexistence of neuropeptides and neurotransmitters

Hökfelt et al. [19] clearly demonstrated the coexistence of neuropeptides and neurotransmitters within the same neuron and sometimes within the same vesicle. While there is little evidence for the coexistence of neuropeptides with the classical neurotransmitters, such as ACh, norepinephrine (NE), dopamine (DA), or serotonin (5HT), two or more neuropeptides may be present in the same or separate secretory granules depending on the processing of the prohormone. These variations permit immense flexibility to the regulation and modification of synaptic events as one transmitter may act on one or several types of postsynaptic receptor; on a presynaptic receptor to inhibit its own release; it may facilitate or inhibit the release of the neurotransmitter. Further possible variations are discussed by Lundberg and Hökfelt [20].

There is considerable evidence that there is selective release of small and large vesicles depending on the frequency and duration of nerve stimulation. In general, peptide release requires bursts of high frequency stimulation but in some cases it is the continuous repetitive stimulation rather than a bursting pattern that regulates peptide release [21]. The neuropeptide is usually responsible for a slow and long-lasting response in contrast to the rapid and short-lived effects of classical transmitters such as ACh and NE.

3.3 Pathway-specific distribution

There appears to be some pattern to the coexistence of neuropeptides according to the neuronal pathways in which they are found. Dorsal root ganglia

(DRG) contain several populations of neurons, each of which has a distinct combination of neuropeptides and a specific projection to its target organ. Some DRG neurons possess SP, calcitonin gene-related peptide (CGRP), CCK and dynorphin, and these neurons project to small blood vessels in the skin. Other DRG neurons contain the first three neuropeptides but not dynorphin, and these neurons project to small blood vessels in skeletal muscle. Yet another group of DRG neurons contain SP, CGRP and dynorphin but not CCK, and innervate the pelvic viscera. Similar pathway-specific neuropeptide combinations occur in sympathetic ganglia.

In contrast, all neurons in some peripheral ganglia have the same neuropeptide combination and project to the same tissue, for example all the cholinergic ciliary ganglion cells contain SP as well as ACh and they all innervate the muscles of the iris of the eye. Gibbins [22] discusses the relationship between co-existence and co-function. The intricacies of neuropeptide regulation may be appreciated by this example: the peripheral axons of unmyelinated sensory nerves release SP, CRGP and neurokinin A (NKA) that synergistically act to reduce inflammation. SP may further enhance wound healing by increasing the rate of fibroblast division. Since CGRP inhibits SP breakdown, both centrally and peripherally, and CCK and dynorphin regulate these effects, the difficulty of clinical administration of agonists or antagonists to individual neuropeptides becomes apparent.

4   Neuropeptide receptors

For neuropeptides to affect their target cells they must bind to cell surface receptor. They may bind to a variety of receptor types and subtypes, with different degrees of affinity. They are characterized by their affinity for the ligand, reversibility, specificity, number and saturation. Desensitization of a receptor may occur after prolonged neuropeptide stimulation and may be due to downregulation of receptor number. The many different isoreceptors that form receptor subtypes can be distinguished mainly by specific drug agonists and antagonists, the use of which experimentally and clinically has proved to be invaluable. Receptor subtypes are exemplified by the arginine vasopressin (AVP) V1 and V2 receptors, which are differentially distributed in tissues and use different second messengers. Five subtypes of the melanocortin receptors have been cloned and the markedly different effects of the melanocortin peptides on the adrenal cortex, melanoma cells, CNS and peripheral neurons are probably due to differential distribution of the receptors in these tissues.

More than 80% of neuropeptide receptors are coupled to G-proteins and stimulate adenylate cyclase and cAMP formation. These receptors are characterized by a core of seven membrane-spanning helices, connected by alternating cytoplasmic and extracytoplasmic loops with a central pore exposed to the extracellular surface.

Other neuropeptides, such as TRH, VP, bombesin and gonadotropin-releasing hormone (GnRH), act through the phospholipase-phosphatidylinositol pathway, involving a different group of second messengers. Some G-protein coupled receptors may activate both types of second messengers and there may be cross-talk between these different signaling systems. A further group of neuropeptide receptors, including atrial natriuretic hormone (ANH) is coupled to guanylate cyclase (cGMP) and yet another set of neuropeptides (insulin, insulin-like growth factor and several other growth factors) couples with tyrosine kinase. The latter 2 groups of receptors have only one membrane-spanning helix. Cytokine receptors are glycoproteins which are classified into several families, only one of which (Type I) includes neuropeptides, such as growth hormone (GH) and prolactin (PRL) receptors.

Fuxe et al. [23] suggest that receptor diversity allows the receptors to couple to different types of G-proteins, thus permitting the stimulation or inhibition of multiple transduction mechanisms, perhaps accounting for different gradations of responses in different tissues. It is obvious that the intrasignaling network of the cell is extraordinarily complex, with ligands showing extensive redundancy, pleitropy, and cross-talk between different signaling systems. The Ca^{2+} sensitivity or insensitivity of various isoforms of receptors and second messenger systems are also important variables. Again, the complexity of receptors, their subtypes and transduction mechanisms provide both a promising yet frustrating field for drug targeting and successful clinical applications.

5   Pharmaceutical potential in treating CNS disorders: genomic, translational and combinational research

It appears that most human disorders have a genetic basis. A large number of genes involved in a single gene disorder have been identified. Consequently the failure of attempts to alleviate human disorders by the administration or inactivation by antagonists of a single neuropeptide is not surprising. Drug metabolism and susceptibility to drug toxicity are also governed by simple genetic changes. While genetic variations may be increase resistance to a disease, treatment for one disease may make an individual more susceptible to other diseases. A combination of genetic and genomic testing of individual patients may be an essential element in tailoring neuropeptide therapy to target a particular disorder in a specific patient [24].

The mapping of the complete sequence of the human genome forms a starting point for making new drugs with the aim of identifying a ligand as a goal for every receptor. However, it must be appreciated that many of the 35,000 to 45,000 genes are involved in common diseases and the science of physiological genomics, which correlates detailed phenotypes to genotypes in patients, is still in its infancy. The innovative technique of tissue-specific, or time-specific, gene knockouts that excise a targeted gene from the genome in the whole animal, will aid dramatically in this search. To have a drug targeted at specific neuropeptide receptor would greatly expand the pharmaceutical potential of treating CNS disorders. Pharmaceutical companies presently are concentrating on translational studies from basic animal research studies to clinical investigations using non-invasive techniques, as well as utilizing combinations of various selected neuropeptides more likely to bring significant clinical effects than individual neuropeptides. These combinational studies are, of course, fraught with problems, not the least of which is gaining approval of the Food and Drug Administration (FDA).

The following sections will summarize the clinical effects presently reported following neuropeptide administration or inactivation. It cannot be expected that a single neuropeptide will cure a single disease and consequently the discussion has been organized according to disorders rather than according to specific neuropeptides.

5.1 Energy homeostasis

Fuel stores and energy metabolism are delicately controlled at optimum levels when homeostasis is achieved. This balance is finely controlled though a complex network of regulatory pathways involving not only the sympathetic nervous system but a variety of neuropeptides and hormones. The hypothalamus is the integrative site for the variety of afferent inputs, not only from circulating neuropeptides and hormones, but also from the gastrointestinal tract and the environment. Plasma concentrations of nutrients are also important regulators of hypothalamic activity. In response, the hypothalamus coordinates the neuropeptide and behavioral responses.

The recent literature on neuropeptides is heavily concentrated on drugs that affect obesity, a condition responsible for many serious complications in humans. The prevalence of obesity, especially in developed countries, has risen sharply and much basic and clinical research has been focused on this disorder. Several genes for monogenic murine obesity have been identified, one of which encodes the hormone leptin that regulates several neuropeptides involved in feeding. In humans, a mutation of the melanocortin-3 receptor gene may be associated with obesity and early onset type 2 diabetes [25]. Eating disorders such as anorexia nervosa and bulimia, while affecting a relatively small population, are serious diseases that may have a neuropeptide as well as a behavioral component, so that appetite-stimulating drugs such as ghrelin are currently of considerable interest.

Many neuropeptides are involved to varying degrees in the control of appetite. Appetite-stimulating (orexigenic) neuropeptides include neuropeptide Y (NPY), ghrelin, melanin-concentrating hormone (MCH), hypocretins/orexins, GH and agouti-related protein (AgRP). Appetite-suppressing (anorectic) neuropeptides include peptide YY (PYY), ST, MSH, galanin, CRH, CCK, glucagon-like peptide 1 (GLP 1), NT and bombesin. Leptin, a hormone secreted by adipose tissue, modifies appetite through its effects on the hypothalamus, and thus on hypothalamic and anterior pituitary neuropeptides. In the following discussion, those substances that appear to be most relevant to clinical tests are discussed first and in more detail.

5.1.1 Appetite-stimulating neuropeptides

Ghrelin is a recently discovered neuropeptide that is a regulator of GH secretion and energy homeostasis. It is an acetylated 28 residue peptide that is the ligand for GH receptors and different GH receptor subtypes, as well as for cells in specific hypothalamic nuclei and in the anterior pituitary hypothalamus. Both rat and human ghrelin and its receptors in the hypothalamus and anterior pituitary have been isolated and cloned [26]. Ghrelin is produced chiefly by the stomach but also by the intestine, placenta, pituitary and possibly by the hypothalamus. It induces hunger in both animals and humans. In humans, circulating ghrelin levels are decreased in chronic obesity and states of positive energy balance, whereas ghrelin levels are increased in patients with anorexia nervosa. A pre-meal rise in human plasma levels suggests a possible role of ghrelin as a hunger signal initiating food ingestion. Plasma ghrelin is influenced by nutritional status and appears to have a reciprocal relationship with plasma insulin, high levels of which suppress ghrelin secretion [27]. In rodents exogenous ghrelin reduces fat utilization, thereby increasing fat deposition, however an adipogenic effect in humans has not yet been demonstrated.

Ghrelin is released from the stomach into the circulation and acts on neuroendocrine networks within the CNS, specifically targeting nuclei in the hypothalamus and AgRP/NPY and POMC cells in the anterior pituitary. These and many other neural and hormonal circuits in the hypothalamus and brain stem are described in detail in an excellent review article by Horvath et al. [28]. Proof that ghrelin crosses the blood-brain barrier was recently provided by Banks et al. [29] in studies that also showed that human ghrelin enters the brain more rapidly than mouse ghrelin. The orexigenic effect of ghrelin is separate from its ability to stimulate the release of GH. It exerts a strong stimulatory effect on GH secretion in humans, releasing more GH than GHRH does, but with which it has a synergistic effect. Ghrelin also stimulates lactotroph and corticotroph secretion [30]. The therapeutic approach of administering ghrelin and/or the unnatural GH-releasing peptide (GHRP) or the combination of one of these peptides with GH, could be a model for the restoration of the physiological secretion of GH and insulin-like growth factor (IGF-I) to normal in short-statured children [31].

NPY is a member of the neuropeptide family that includes pancreatic polypeptide (PPP) and PYY. NPY is the most abundant neuropeptide in the rat brain and is found in the periphery in nerve plexuses surrounding blood vessels in a variety of organs. NPY administered centrally results in marked stimulation of feeding in satiated rats whereas food deprivation increases NPY hypothalamic levels. Chronic infusion of NPY results in obese animals. These actions of NPY are exerted through NPY receptor subtypes as demonstrated by pharmacological antisense "knockdown" and targeted gene "knockout" approaches. Few clinical studies have been reported utilizing specific receptor antagonists to NPY. This topic is reviewed by Gehlert [32].

Orexin is an orexigenic neuropeptide produced primarily in the lateral hypothalamus, a region implicated in driving feeding. These orexin neurons project widely in the brain and receive terminal appositions from NPY-,

AgRP- and α-MSH-IR fibers. Orexin neurons are innervated by peptidergic fibers corresponding to leptin-responsive cell types from the arcuate nucleus and thus may link peripheral metabolic cues to autonomic regulatory sites and neuroendocrine mechanisms [33]. Important regulatory signals may be a fall in plasma glucose (stimulatory) countered by satiety signals generated by eating, such as gastric distention (inhibitory). Orexins appear to stimulate food intake in the short term by recruiting different neural circuits and exert different effects on food choice [34]. Antipsychotic drugs are associated with weight gain and there is evidence that the drug-induced weight gain is associated with the activation of distinct orexin neurons innervating the frontal cortex [35]. While orexins are involved in the stimulation of food intake, they also may modulate central nervous control of arousal and sleep-wake mechanisms [36, 37].

Galanin is a 29-amino acid neuropeptide not homologous with any other known peptide. High concentrations of galanin and its receptor subtypes are present in the hypothalamus and central administration of the neuropeptide increases food consumption in fed rats, a highly reproducible result. Although galanin appears to play a minor role in normal feeding behavior it is possible that under abnormal physiological conditions it may exert profound effects. Preliminary reports demonstrate unusual endogenous peptide levels in human appetite disorders but other studies have been inconclusive. It is possible that more targeted investigations involving subtype-selective non-peptide galanin receptor ligands may have clinical implications in appetite disorders [38].

MCH is a neuropeptide that has antagonistic actions to those of MSH, both in pigment dispersal and in feeding behavior [39]. MCH is an orexigenic peptide produced in the lateral hypothalamus, and which, when administered intracerebroventricularly interacts with a number of other peptides that affect appetite. GLP-1 and NT both inhibit the ability of MCH to induce feeding. Ablation of the MCH gene in mice results in a marked weight loss, with corresponding loss of body fat and decreased leptin levels.

AgRP is a 131 amino-acid protein that acts as a high-affinity antagonist of α-MSH on the melanocortin MC1 and MC4 receptor level. It antagonizes the synthesis of black pigment normally stimulated by α-MSH and causes obesity in mice suffering from autosomal mutations. These mice also are characterized by yellow hair as a result of the overproduction of the agouti protein. The current evidence indicates a dual-appetite regulating system consisting of POMC and AgRP/NPY expressing neurons in the arcuate hypothalamic nucleus. These neurons project to other hypothalamic nuclei to widen their influence on autonomic centers and other neuropeptides affecting feeding behavior [40]. Elucidation of these complex neuronal circuits, together with creative pharmacological tools, is essential for the success of clinical applications. Certainly appetite-enhancing neuropeptides should be investigated for the treatment of the anorexia of cancer patients.

5.1.2 Appetite-inhibiting neuropeptides

Peripheral satiety systems include PYY, CCK, bombesin-like peptides, ST, GLP-1, insulin, amylin and NT. These peptides are released by the gastrointestinal tract during a meal and act as negative-feedback control of the size of the meal [41]. Many of these peptides reach the brain via the circulation, whereas others involve afferent vagal fibers. Analyses of the behavioral, physiological and neural mechanisms of these hormones indicate that each is involved in the direct or indirect control of meal size. It is still not clear what their interaction is in normal eating or in eating disorders, complex problems that have to be clarified for their full therapeutic potential to be realized.

Hypothalamic neuropeptides involved in appetite suppression include CRH and its homologue urocortin, galanin, orexin, CART and melanocortins. Melanocortins are short amino acid sequences derived from the 1-13 amino acid sequence of α-MSH. This sequence is identical to the 1-13 sequence within the ACTH 1-39 molecule. These peptide fragments, unlike the parent molecule ACTH 1-39, do not stimulate the adrenal cortex but have significant effects on both central and peripheral neurons. The most potent of these peptide fragments are ACTH 4-9, ACTH 4-10, an ACTH 4-9 analog, ORG 2766 and an ACTH 4-10 analog BIM 22015 [42].

One of the most effective weight-loss drugs is D-fenfluramine (d-FEN) but it was withdrawn from clinical use because of reports of cardiac complications in a subset of patients. In rodents d-FEN's anorexic actions requires activation of CNS melanocortin pathways, indicating that other drugs targeting these pathways may provide more effective, selective and therefore safer anti-obesity treatment [43].

Both the appetite-suppressing neuropeptides that inhibit feeding behavior to different degrees and drugs that enhance these satiety systems are suitable clinical candidates for the control of obesity. It is beyond the scope of this review to discuss all these neuropeptides: the emphasis here is on those neuropeptides that are currently the focus of intensive investigation for their clinical potential.

Briefly, the anorexic effects of the hypothalamic neuropeptide, CRH/urocortin, are linked to the well-known stress response that inhibits sexual activity and food intake [44]. More direct therapeutic possibilities arise from a recent report that indicates that the intestinal peptide PYY has clinical applications in the control of hunger. PYY is a member of the neuropeptide Y family from which the mammalian PPP probably arose by gene duplication [45, 46]. PYY reaches the hypothalamus via the circulation, acting directly on feeding centers in the arcuate nucleus that coordinate NPY and melanocortin neural circuits. The long-lasting effect of PYY is unlike that of other gut-derived signals, such as CCK, which are considered to be "short-term" satiety signals [47]. Administration of a physiological dose of PYY, similar to that physiologically released after a normal meal, decreases appetite and reduces food intake by 33% over 24 hours [48]. This preliminary study involved only 12 subjects, none of whom was obese. It remains to be tested whether these important effects are to be obtained in obese individuals. Like other peptides, PYY has to be administered by injection.

Somatostatin is produced both peripherally in the gut and centrally in the hypothalamus and may be considered to be a general inhibitor since many other neuropeptides evoke their effects indirectly through stimulation of ST. The long-acting stable analog of ST, octreotide, has a number of therapeutic indications. It is effective in treating acromegaly, as well as many gastrointestinal disorders and new analogs and delivery systems make it even more stable and long-lasting [49].

5.1.3 Integration of neuropeptides affecting appetite by leptin

Leptin, a satiety signal, is released primarily from white adipose tissue but may also be produced by other tissues including the anterior pituitary gland [50]. It is a 16 kDa protein that interacts with leptin receptors in the brain and in many other tissues and is thought to regulate energy balance through effects on food intake and thermogenesis. Leptin may mediate the neuroendocrine response to starvation and modulate the stress response and timing of puberty. Flier and his co-workers have shown that leptin may also be involved in the maturation and function of the neurendocrine axis [51, 52].

Leptin regulates the expression of several neuropeptide genes, including POMC, NPY and AgRP [53, 54]. It is an important regulator of many neuroendocrine systems affecting body weight.

A loop system exists between the peripherally secreted hormone leptin and the hypothalamic neuropeptides α-MSH and NPY. Leptin stimulates the hypothalamic expression of α-MSH, an orexigenic peptide, while inhibiting the expression of MCH, an anorexigenic peptide, actions which reflect the opposing effects of these two neuropeptides on animal pigmentation [39, 55]. When hypothalamic NPY levels are increased by fasting or by i.c.v infusion, food intake and body weight gain increase. A feedback system between leptin and NPY also exists. NPY increases insulin and corticosterone levels which, in turn, result in lipolytic activity in adipose tissue and leptin release.

Interactions between leptin and PRL-releasing peptide (PrRp) have also been reported [56]. PrRp is a novel anorexigen that reduces food intake and prevents body weight gain in rats. Immunohistochemical studies show that PrRP neurons contain leptin receptors, indicating that PrRp neurons may form part of a leptin-sensitive brain circuitry involved in the regulation of food intake and energy homeostasis.

The many facets of leptin action, apart from its anti-obesity role through inhibition of food intake through actions on hypothalamic centers, have been challenged by Unger [57]. Unger presents evidence for an alternative physiological role: an antisteatotic activity in which fatty acid overaccumulation in nonadipose tissues is prevented by leptin-mediated regulation of β-oxidation. While leptin acts centrally on the hypothalamus in lean or on leptin-deficient animals, when hyperleptinemia exceeds 15ng/ml, as in obesity, a further rise in plasma leptin levels does nor raise cerebrospinal leptin levels or reduce food intake. In these cases, the peripheral antisteatotic action is maintained. This suggests that at chronically hyperleptinemic levels the hormone acts chiefly on peripheral tissues and its hypothalamic action has reached a plateau. This observation would account for the leptin resistance of obese humans to administration of leptin. Similarly, the central targets of leptin that mediate the transition from starvation to the fed state may be distinct from those that mediate the response to overfeeding and obesity [58].

Due to leptin's influence on the anterior pituitary gland, this hormone also affects the hypothalamic-anterior pituitary-adrenal axis (HPA), the hypothalamic-anterior pituitary-thyroid axis and the hypothalamic-anterior pituitary-gonadal axis. Leptin deficiency leads to morbid obesity in experimental animals and narcoleptic humans, who also have disruptions in their sleep-wake cycle [59]. These observations lend credence to the belief that leptin has many functions apart from its important role as a regulator of body weight and energy homeostasis.

Leptin, its analogs and antagonists are currently the focus of active clinical trials. Unfortunately, the dramatic appetite-suppressing action of leptin in animals is not seen in obese humans. This is not unexpected in that a wide variety of biological and behavioral phenomena are involved in feeding. Clinically effective anti-obesity drugs may act on more than one component of the appetite system, or a suitable combination of drugs, together with alterations in diet may be needed for successful control of the obesity epidemic in developed countries.

5.2   Neuropeptides affecting CNS behavioral disorders

The pioneering studies of De Wied [60] and of Kastin and his colleagues [61] have shown that neuropeptides derived from POMC and the neurohypohyseal hormones OT and VP have profound effects on learning and memory processes, grooming, stretching and yawning, social, sexual and rewarded behavior, and memory, in rodents. This vast body of work has been reviewed recently [62]. Both animal and clinical studies suggest that these neuropeptides may be beneficial in aging, neuropathy, memory disturbances, pain, depression, impotence, obesity, fever and insomnia [63].

5.2.1 Intractable seizures

Neuropeptides are also implicated in therapy for intractable seizures. Despite current advances in the treatment of epilepsy, these seizures remain a significant therapeutic challenge. A review article by Kubek and Garg [64] cites evidence that TRH and selected TRH analogs have antiepileptic effects in several models of animal seizure, including kindling and electroconvulsive shock. Clinically, TRH treatment has been reported to be effective in intractable epilepsies such as infantile spasms. Infantile seizures are age-specific and are associated with mental retardation. Conventional anticonvulsants have little effect but treatment with high dosages of ACTH arrests the seizure completely. However, the severe side effects of the neuropeptide are a severe drawback to this therapy. Baram and Hatalski [65] review evidence that CRH is an important mechanism for the generation of developmentally regulated, triggered seizures. As ACTH suppresses the secretion of CRH, a natural convulsant, by negative feedback, a trial of 6 infants with infantile spasms non-responsive to other treatments was established. These infants received α-helical CRH, a competitive antagonist of the peptide. This treatment was ineffective and the authors conclude that since no central effects on arousal, seizures or EEG were observed, this CRH antagonist did not cross the blood-brain barrier. They suggest that nonpeptide compounds that reach CNS receptors are required to check the hypothesis that blocking CRH receptors may ameliorate infantile spasms and mental retardation [66].

5.2.2 Alzheimer's disease

Marked reductions in ST-IR and CRH-IR are prominent neurochemical deficits in Alzheimer's disease (AD). In a postmortem study of 66 elderly patients, both ST-IR and CRH-IR were reduced in patients with severe dementia, but only CRH was reduced significantly in the cortices of those with mild dementia. Thus CRH-IR can serve as a potential neurochemical marker of early dementia and possibly early AD [67].

A comprehensive hypothesis for an understanding of the pathophysiological changes that lead to the clinical manifestation of AD has been propounded by Heininger [68]. He suggests that the delicate network that regulates the homeostatic balance of neurotrophic factors (NPY, ST, SP, GH, insulin and insulin-like growth factors, together with gonadal hormones), is disturbed by aging, leading to a hyper-responsiveness to stress and loss of the diurnal rhythm of glucocorticoids. Neurotransmitters such as serotonin and dopamine are also intimately involved in the aging process. The imbalance between neuroprotective and neuroaggressive factors leads to changes that are similar to those of AD and permit the development of the disease. This holistic approach is daunting when applied to clinical application as no single therapy can hope to alleviate the devastating effects of this disease.

A more encouraging view of the clinical potential of pharmaceutical research is presented by Oliver et al. [69]. They argue that neuropeptide systems offer certain characteristics that distinguish them from neurotransmitters and thereby make neuropeptides attractive for drug targeting. This is especially relevant for identified antagonists of neuropeptide receptors. Information derived from bioinformatics, proteomic and transgenic approaches should ultimately lead to clinical benefits.

Treatment of AD traditionally involves enhancement of cholinergic transmission through adminstration of acetylcholinesterase inhibitors and new cholinergic agonists and enhancers are currently in development. A review by Emre and Qizilbash [70] cites other therapeutic approaches directed toward modulation of neurotransmitters, neuropeptides and those substances acting via excitatory amino acid receptors, such as ampakines or N-methyl-D-aspartate (NMDA) antagonists. Introduction of atypical neuroleptics is one of the more recent developments in the treatment of behavioral symptoms. Other approaches attempt to decrease the accumulation of β-amyloid. An exciting development is vaccination with amyloid-β peptide, a vaccine that appears to be effective in an animal model and is currently in clinical trials. The possible protective action of estrogens is also a controversial topic at the moment.

5.2.3 Depression

While animal studies demonstrate that neurotransmitter systems are involved in depression, the neuropeptides SP, CRH, NPY, VP and ST are also important factors [71]. An SP antagonist has been shown to have clinical efficiency in depression [72]. The therapeutic use of antidepressants affecting serotonin transmission is widespread but these are characterized by slow onset of action and side effects. Holsboer [73] suggests that the corticosteroid receptor hypothesis of depression leads to a new drug target, that of CRH receptors, and he discusses new strategies for screening compounds that include the use of DNA microarrays, searches of compound libraries to behavioral screens of mouse mutants. Patients with stress-associated disorders such as depression or intensive anxiety have elevated levels of CRH and dynamic endocrine studies indicate that both hypothalamic and extra-hypothalamic concentrations of CRH are increased. Clinical studies show that CRH is the principal neuropeptide involved in regulating stress [74]. While the causes of the increased CRH in depression are unknown, the involvement of glucocorticoid receptors as a result of stressors is likely. Clinically effective antidepressant therapies often normalize CRH, and the dampening of hyperactivity of the HPA system by antidepressants is associated with successful treatment of psychopathology [75]. Mitchell [76] suggests that the careful manipulation of CRH may hold therapeutic promise for sufferers of mood disorders. A review of a possible significant role for arginine VP in the pathophysiology of major depression proposes that future antidepressants may target the vasopressinergic system as AVP and CRH act synergistically to stimulate ACTH release from the pituitary [77].

5.2.4 Sleep disorders

Various neuropeptides regulate the electrophysiological and neuroendocrine control of sleep. The balance between GHRH and CRH plays a key role in normal and pathological sleep regulation. In normal young subjects, GHRH stimulates slow-wave sleep and GH secretion but inhibits cortisol release, whereas CRH has the opposite effect. In normal elderly subjects and in patients with acute depression, the GHRH ratio is changed in favor of CRH, resulting in disturbances in sleep endocrine activity. Galanin and NPY promote sleep, whereas in the elderly ST impairs sleep. Rapid eye movement is modulated by VIP. The impact of delta sleep-inducing peptide on human sleep regulation is not yet clear [78, 79]. Sleep disturbances are characteristic of normal aging and no long-term drugs that promote sleep without inducing addiction are available. Thus investigation of neuropeptide actions on sleep should be a profitable guide for investigations leading to clinical application.

5.2.5 Obsessive compulsive disorder

Present therapy for obsessive compulsive disorder (OCD) centers on antidepressants, of which serotonin uptake inhibitors are the most potent. However, the extensive interaction between brain neuropeptidergic and monoaminergic systems lends credibility to the belief that a wide variety of neuropeptides may be involved. In a review by McDougle et al. [80] the role of different neuropeptides in the clinical neurobiology of children, adolescents and adults with OCD is discussed, emphasizing the importance of the development of age-related symptoms. The recent discovery of novel drugs, termed peptoids, that mimic or block neuropeptide function may permit a high degree of selectivity of drug action. Peptoid agonists can exert extremely powerful actions on brain function. Peptoid antagonists appear to be relatively free of side effects since neuropeptide systems are only activated under very selective conditions. High efficiency kappa opioid receptor agonists such as CI-977 (enoline) have potential for the treatment of pain and stroke while highly selective CCK-B antagonists such as CI-988 are helping to clarify the mechanisms underlying the treatment of anxiety disorders and drug abuse.

5.3   Neuropeptides and inflammation

A recent study of neuroendocrine involvement in acute inflammatory episodes uses leprosy as an example of the interaction of peripheral and CNS factors [81]. Multiple components of the neuroendocrine system, both local and systemic are involved in the regulation of inflammation. Peptidergic sensory fibers not only signal the CNS but also release active peptides into the tissues that they innervate. These include the tachykinins (SP and NKA), CGRP and VIP. Most of these neuropeptides are proinflammatory. T cells of the immune system bear receptors for these peptides and can dramatically affect cytokine production. Cytokines, in turn, activate the sympathetic nervous system and the HPA axis, resulting in release of cortisol, a major anti-inflammatory hormone. Cortisone and cortisone analogs, such as prednisolone, are among the most effective treatments for leprosy pathology [82]. In leprosy sensory C fibers and sympathetic innervation are destroyed and anti-inflammatory feedback circuits involving the HPA axis are blunted, aggravating the inflammation. Thus communication between the nervous, endocrine and immune systems is essential for the maintenance of physiological equilibrium. ST has been proposed as the intersystem signaling molecule, with emphasis on the immune system. Systemic or local treatment with ST or ST analogs is beneficial in animal models of autoimmune disease and chronic inflammation, apparently antagonizing the effects of SP [83].

α-MSH is one of the principal neuroimmunomodulating neuropeptides and seems to exercise some control on the cutaneous inflammatory process. This appears to be a bi-directional circuit between the brain, immune system and the skin. This process involves both a central action. mediated by descending anti-inflammatory pathways and a local direct action on inflammatory cells infiltrating the dermis. The production of inflammatory cytokines by cells invading the dermis, such as monocytes, macrophages and neutrophils, is down-regulated by α-MSH while that of the anti-inflammatory cytokine IL-10 is stimulated [84]. In a study of patients with infectious and inflammatory disease, it was found that levels of α-MSH were elevated in HIV-infected patients and in the synovial fluid of arthritis patients. Treatment of HIV-infected patients and patients with septic syndrome with α-MSH reduced production of cytokines in whole blood samples stimulated with endotoxin.

The involvement of the inflammatory system in evoking the secretion of neuropeptides and proinflammatory cytokines in the syndrome of patients with complex regional pain, is extensively discussed by Huygen et al. [85]. The possible development and application of drugs that could act through selective receptor antagonism or inhibition of enzymatic synthesis may prevent further stimulation of this cascade. α-MSH may play a role in pain produced by inflammation as it alleviates neuropathic pain, both acute and chronic, in rats following chronic infusion of the melanocortin-receptor antagonist SHU9119. This effect is most likely mediated through the MC4 receptor in the spinal cord [86]. These several central and peripheral effects of α-MSH indicate that it might be worthwhile to investigate its clinical possibilities for the treatment of inflammatory disease in humans [87] especially as many clinical studies have shown this neuropeptide to be well-tolerated.

5.4 Neuroregenerative and neuroprotective action of neuropeptides

Growth factors are implicated in the development and regeneration of both central and peripheral neurons and it is beyond the scope of this review to enumerate the many review articles on this topic. Instead I have chosen to emphasize the role of the melanocortins since this is the area with which I am most familiar.

There is extensive evidence from animal models that melanocortins accelerate recovery from nerve trauma, ameliorate many neuropathies and may be helpful in alleviating convulsive seizures. The marked effects of the melanocortins on the development and regeneration of peripheral nerves in rats is described in several reviews [88–90] but there is also considerable evidence for a significant effect on CNS neurons. In tissue culture studies, melanocortins have been shown to exert a direct effect on cultured primary sensory and motor neurons. Neurons from different CNS regions respond differently to specific melanocortins [91–93].

In the intact animal the melanocortins act centrally, when administered prenatally or perinatally, to accelerate the integrated development of the neuromuscular system [42, 94, 95] and to affect the development of sexual behavior [96, 97]. The poor regenerative properties of CNS neurons can be ameliorated to some extent by modifying their environment. The melanocortins are one of the many trophic factors that appear to be moderately successful in ameliorating the deleterious effects of brain or spinal cord damage. Various peptide fragments of ACTH, referred to collectively as melanocortins, have been administered to animals with different brain lesions (neocortical, hippocampal and sham lesions) and the results on cognitive behavior and learning evaluated. The results have been extremely mixed. Part of the problem appears to be variation in the specific melanocortin peptide administered, its timing and dosage. Compensation, perhaps including collateral sprouting of remaining inputs and the subsequent formation of new synapses, may play an important part in recovery from brain damage [98–102]. More consistent evidence for a positive role of the melanocortins in recovery from brain lesions comes from experiments in which recovery of motor function can be quantitatively assayed. In a model involving lesions of the nigrostriatal system, a dopaminergic system involved in Parkinson's disease and other motor dysfunctions in humans, administration of a synthetic melanocortin, ORG 2766, an ACTH 4-9 analogue, accelerated spontaneous recovery [103]. Postsynaptic sensitivity of the dopaminergic receptors occurs following unilateral nigrostriatal lesions and the resulting denervation supersensivity is accelerated in the peptide treated rats. The rapidity with which this occurs favors increased neuronal survival rather than regeneration, indicating a neuroprotective or compensatory role for this peptide. In addition, ORG 2766 acutely enhances morphological and biochemical recovery following nigrostriatal destruction [104].

Clinically, successful ACTH administration to patients with a deficiency of this anterior pituitary hormone, is well-established but reports on the clinical administration of its non-corticosteroidogenic fragments for disorders of the CNS are much more limited. Clinical trials involving the administration of ORG 2766 to prevent the development of neuropathy in women with ovarian cancer treated with cisplatin have had positive results [105, 106] although extensive Phase 3 trials using vibration threshold as an indicator of neuropathy were discontinued. Some encouraging results were obtained in a small pilot study of patients with progressive spinal muscular atrophy, carpal tunnel syndrome and multiple sclerosis [107]. Diseases primarily of myogenic origin, such as muscular dystrophy, do not respond to this peptide.

6 Summary

The general characteristics of neuropeptides are discussed as a background for the understanding of their role in regulation of physiological systems. The extent of those systems that are crucially affected by neuropeptides is vast and the complexity of their interactions makes the clinical focus on a specific neuropeptide unsatisfactory. The clinical potential of neuropeptides affecting eating disorders, CNS behavioral disorders and the neuroregenerative and neuroprotective action of neuropeptides is discussed. It is probable that successful neuropeptide therapeutics will depend upon the application of translational and combinational research using various ingenious combinations of neuropeptides, their agonists and antagonists, neuropeptide receptor agonists and antagonists, improved methods of delivery and the development of peptides targeted to the genetic profile of individual patients.

References

1   DeWied D (1969) Effects of peptide hormones on behavior. In: WF Ganong, L Martini (eds): *Frontiers in Neuroendocrinology*. Oxford University Press, New York, 97–140

2   Sandman CA, Schally AV, Kastin AJ, Miller L H (1972) A neuroendocrine influence on attention and memory. *J Comp Physiol Psychol* 80: 54–58

3   Kastin AJ, Olson RD, Schally A V, Coy DH (1979) CNS effects of peripherally administered brain peptides. *Life Sci* 25: 401–414

4   Strand FL, Saint-Come C, Lee TS, Lee SJ, Kume JA, Zuccarelli LA (1993) An ACTH/MSH 4-10 analog BIM 22015 has neurotrophic and myotrophic attributes during peripheral nerve regeneration. *Peptides* 14: 287–296

5   Strand FL (1999) *Neuropeptides: Regulators of Physiological Processes*. The MIT Press, Cambridge, MA, 11–12

6   Strand FL, Segarra AC, Zuccarelli LA, Kume J, Rose KJ (1990) Neuropeptides as neuronal growth regulating factors. *Ann NY Acad Sci* 579: 68–80

7   Hökfelt T, Zhang X, Wiesenfeld-Hallin Z (1994) Messenger plasticity in primary sensory neurons following axotomy and its functional implications. *Trends Neurosci* 17: 22–30

8   Stefano GB (1989) Opioid peptides – comparative peripheral mechanisms. In: S Holmgren (ed): *Comparative Physiology of Regulatory Peptides*. Chapman and Hall, New York, 122–129

9   Acher R, Chauvet J (1995) The neurohypophyseal regulatory cascade. *Front Neuroendocrinol* 16: 237–289

10   Acher R (1980) Molecular evolution of biologically active polypeptides. *Proc Soc Lond Biol Sci* 210: 1–43

11   Blobel G, Dobberstein B (1975) Transfer of proteins across membranes. *J Cell Biol* 67: 835–851

12   Mains RE, Cullen EI, May V, Eipper BA (1987) The role of secretory granules in peptide biosynthesis. *Ann NY Acad Sci* 493: 279–291

13 Rouille Y, Duguay S J, Lund K, Furuta M, Gong QM, Lipkind G, Oliva AA, Chan SJ, Steiner DF (1995) Proteolytic processing mechanisms in the biosynthesis of neuroendocrine peptides. *Front Neuroendocrinol* 16: 322–361

14 Nakanishi SY, Inoue I, Kita T, Nakamura M, Chang ACY, Cohen SN, Numa S (1979) Nucleotide sequence of cloned DNA for bovine corticotropin/β-lipotropin precursor. *Nature* 278: 423–429

15 Nakanishi S, Teranishi Y, Wananabe Y, Notake M, Noda M, Kakidani H, Jingami H, Numa S (1981) Isolation and characterization of the bovine corticotropin/β-lipotropin precursor gene. *Eur J Biochem* 115: 429–438

16 Schlegel W, Mollard P (1995) Electrical activity and stimulus-secretion coupling in neuroendocrine cells. In: H Scherübl, H Hescheler (eds): *The Electrophysiology of Neuroendocrine Cells*. CRC Press, Boca Raton, Florida, 23–38

17 Nicholls DG (1994) *Proteins, Transmitters and Synapses*. Blackwell, Oxford, 142–152

18 Burgoyne RD, Morgan A (1995) Ca^{2+} and secretory-vesicle dynamics. *Trends Neurosci* 18: 191–196

19 Hökfelt T, Holets VR, Staines W, Meister B, Melander T, Schalling M, Schultzberg M, Freedman J, Bjorklundh H, Olson L, Lindh B et al (1986) Coexistence of neuronal messengers – an overview. *Prog Brain Res* 68: 33–70

20 Lundberg JM, Hökfelt T (1985) Coexistence of peptides and classical neurotransmitters. In: Neurotransmitters in action. *Prog Brain Res* 68: 33–70

21 Peng Y, Horn JP (1991) Continuous repetitive stimuli are more effective than bursts for evoking LHRH release in bullfrog sympathetic ganglia. *J Neurosci* 11: 85–95

22 Gibbins IL (1989) Co-existence and co-function. In: S Holmgren (ed): *The Comparative Physiology of Regulatory Peptides*. Chapman and Hall, New York, 308–343

23 Fuxe K, Li X-M, Tanganelli S, Hedlund P, Oconnor WT, Ferraro L, Ungerstedt U, Agnati LF (1995) Receptor-receptor interactions and their relevance for receptor diversity-focus on neuropeptide/dopamine interactions. *Ann NY Acad Sci* 757: 365–376

24 Kucherlapati R (2002) Genetics, genomics and the practice of medicine. *Regul Pept* 108: 1 (Abstract)

25 Wong J, Love DR, Kyle C (2002) Melanocortin-3 receptor gene variants in a Maori kindred with obesity and early onset type 2 diabetes. *Diab Res Clin Pract* 58: 61–71

26 Bowers CY (1999) GH releasing peptides (GHRPs). In: J Kostyo, H Goodman (eds) *Handbook of Physiology*. Oxford University Press, New York, 267–297

27 Saad MF, Bernaba B, Hwu CM, Jinagouda S, Fahmi S, Kogosov E, Boyadjian R (2002) Insulin regulates plasma ghrelin concentration. *J Clin Endocrinol Metab* 87: 3997–4000

28 Horvath TL, Diano S, Sotonyi P, Heiman M, Tschop M (2001) Minireview: Ghrelin and the regulation of energy balance – a hypothalamic perspective. *Endocrinology* 142: 4163–4169

29 Banks WA, Tschöp M, Robinson SM, Heiman ML (2002) Extent and direction of ghrelin transport across the blood-brain barrier is determined by its unique structure. *Pharmacol Exper Ther* 302: 822–827

30 Arvat E, Maccario M, Di Vito L, Broglio F, Benso A, Gottero C, Papotti M, Muccioli G, Dieguez C, Casanueva FF, Deghenghi R et al (2001) Endocrine activities of ghrelin, a natural growth hormone secretagogue (GHS), in humans: comparison and interaction with hexarelin, a nonnatural peptidyl GHS, and GH-releasing hormone. *Clin Endocrinol Metab* 86: 1169–1174

31   Bowers CY (2001) Unnatural growth hormone-releasing peptide begets natural ghrelin. *J Clin Endocrinol Metab* 86: 1464–1469

32   Gehlert DR (1999) Role of hypothalamic neuropeptide Y in feeding and obesity. *Neuropeptides* 33: 329–338

33   Sakurai T (1999) Orexins and orexin receptors: implication in feeding behavior. *Regul Pept* 85: 25–30

34   Clegg DJ, Air EL, Woods SC, Seeley RJ (2002) Eating elicited by orexin-a but not by melanin-concentrating hormone, is opioid mediated. *Endocrinol* 143: 2995–3000

35   Fadel J, Bubser M, Deutch AY (2002) Differential activation of orexin neurons by antipsychotic drugs associated with weight gain. *J Neurosci* 22: 6742–6746

36   Sunter D, Morgan I, Edwards CMB, Dakin Cl, Murphy KG, Gardiner J, Taheri S, Rayes E, Bloom SR (2001) Orexins: effects on behavior and localization of orexin receptor 2 messenger ribonucleic acid in the rat brainstem. *Brain Res* 907: 27–34

37   Tahari S, Zeitzer JM, Mignot E (2002) The role of hypocretins (orexins) in sleep regulation and narcolepsy. *Annu Rev Neurosci* 25: 283–3313

38   Crawley JN (1999) The role of galanin in feeding behavior. *Neuropeptides* 33: 369–375

39   Qu D, Ludwig DS, Gammeltoft S, Piper M, Pelleymounter MA, Cullen MJ, Mathes WF, Przypek J, Kanarek R, Maratos-Flier E (1996) A role for melanin-concentrating hormone in the central regulation of feeding behavior. *Nature* 380: 243–247

40   Elmquist JK, Elias CF, Saper CB (1999) From lesions to leptin: hypothalamic control of food intake and body weight. *Neuron* 22: 221–232

41   Gibbs J, Young R C, Smith G P (1973) Cholecystokinin decreases food intake in rats. *J Comp Physiol Psychol* 84: 488–495

42   Strand FL, Alves SE, Antonawich FJ, Lee SJ, Lee TS, Zuccarelli LA (1994) Developing and regenerating systems as models for trophic effects of ACTH neuropeptides. In: T Palomo, T Archer, T Beninger (eds): *Strategies for studying brain disorders: Vol 2. Schizophrenia, movement disorders and age related cognitive disorders.* Farrand Press, London, 389–407

43   Heisler LK, Cowley MA, Tecott LH, Fan W, Low MJ, Smart JL, Rubinstein M, Tatro JB, Marcus JN, Holstege H, Lee CE et al (2002) Activation of central melanocortin pathways by fenfluramine. *Science* 297: 609–611

44   Heinrichs SC, Richard D (1999) The role of corticotropin-releasing factor and urocortin in the modulation of ingestive behavior. *Neuropeptides* 33: 350–359

45   Larhammar D, Blomqvist, Soderberg C (1993) Evolution of the neuropeptide Y family of peptides. In: WF Colmers, C Wahlenstedt (eds): *The biology of neuropeptide Y and related peptides.* Totawa, New Jersey, 1–42

46   Conlon JM (2002) The origin and evolution of peptide YY (PYY) and pancreatic polypeptide (PP). *Peptides* 23: 269–278

47   Schwartz MW, Woods SC, Porte D, Seeley RJ, Baskin DG (2000) Central nervous control of food intake. *Nature* 404: 661–671

48   Batterham RL, Cowley MJ, Small CJ, Herzog H, Cohen MA, Dakin CL, Wren AM, Brynes AE, Low MJ, Ghatei MA, Cone RD et al (2002) Gut hormone PYY (3–36) physiologically inhibits food intake. *Nature* 418: 650–654

49   Edwards CMB, Cohen MA, Bloom SR (1999) Peptides as drugs. *Q J Med* 92: 1–4

50   Lloyd RV, Jin L, Tsumanuma I, Vidal S (2001) Leptin and leptin receptor in anterior pituitary function. *Pituitary* 4: 33–47

51   Ahima RS, Prabakaran D, Mantzoros C, Qu DQ, Lowell B, Maratos-Flier E, Flier JS (1996) Role of leptin in the neuroendocrine response to fasting. *Nature* 382: 250–252

52   Ahima RS, Prabakaran D, Flier JS (1998) Postnatal leptin surge and regulation of circadian rhythm of leptin by feeding. *J Clin Invest* 101: 1020–1027

53   Schwartz MW, Seeley RJ, Woods SC, Weigle DS, Campfield LA, Burn P, Baskin DG 1997) Leptin increases hypothalamic pro-opiomelanocortin mRNA expression in the rostral arcuate nucleus. *Diabetes* 46: 2119–2123

54   Thornton JE, Cheung CC, Clifton DK, Steiner RA (1997) Regulation of hypothalamic proopiomelanocortin mRNA by leptin in ob/ob mice. *Endocrinol* 138: 5063–5066

55   Ludwig DS, Mountjoy KG, Tatro JB, Gillette JA, Frederich RC, Flier JS, Maratos-Flier E (1998) Melanin-concentrating hormone: a functional melanocortin antagonist in the hypothalamus. *Am J Physiol* 274: E627–E633

56   Lawrence CB, Ellacott, KLJ, Luckman SM (2002) PRL-releasing peptide interacts with leptin to reduce food intake and body weight. *Endocrinol* 143: 368–374

57   Unger R H (2000) Leptin physiology: a second look. *Regul Pept* 92: 87–95

58   Ahima RS, Kelly J, Elmquist JK, Flier JS (1999) Distinct physiologic and neuronal changes to decreased leptin and mild hyperleptinemia. *Endocrinol* 140: 4923–4931

59   Kok SW, Meinders AE, Overeem S, Lammers GJ, Roelfsema F, Frolich M, Pijl H (2002) Reduction of leptin levels and loss of its circadian rhythm in hypocretin (orexin) deficient narcoleptic humans. *J Clin Endocrinol Metab* 87: 805–809

60   De Wied D (1964) Influence of anterior pituitary on avoidance learning and escape behavior. *Am J Physiol* 207: 255–259

61   Kastin AJ, Plotnikoff, Schally AV, Sandman CA (1976) Endocrine and CNS effects of hypothalamic peptides and MSH. In: S Ehrenpreis, IJ Kopin (eds): *Reviews of neuroscience*. Raven Press, New York, 111–148

62   De Wied D (1999) Behavioral pharmacology of neuropeptides related to melanocortins and the neurohypophyseal hormones. *Eur J Pharmacol* 375: 1–11

63   Kastin AJ, Ehrensing RH, Banks WA, Zadina JE (1987) Possible therapeutic implications of the effects of some peptides on the brain. *Prog Brain Res* 72: 223–234

64   Kubek M J, Garg BP (2002) Thyrotropin-releasing hormone in the treatment of intractable epilepsy. *Pediatric Neurol* 26: 9–17

65   Baram TZ, Hatalski CG (1998) Neuropeptide-mediated excitability – a key triggering mechanism for seizure generation in the developing brain. *Trends Neurosci* 21: 471–476

66   Baram, TZ, Mitchell WG, Brunson K, Haden E (1999) Infantile spasms – Hypothesis-driven therapy and pilot human infant experiments using corticotropin-releasing hormone-receptor antagonists. *Dev Neurosci* 21: 281–289

67   Davis KL, Mohs RC, Marin DB (1999) Neuropeptide abnormalities in patients with early Alzheimer's disease. *Archiv Gen Psychiatry* 56: 981–987

68   Heininger K (2000) A unifying hypothesis of Alzheimer's disease. *Rev Neurosci* 11: 213–328

69   Oliver KR, Sirinathsinghji DJS, Hill RG (2000) From basic research on neuropeptide receptors to clinical benefit. *Drug News Perspect* 13: 530–542

70   Emre M, Qizilbash N (2001) Experimental approaches and drugs in development for the treatment of dementia. *Expert Opinion Investig Drugs* 10: 607–617

71   Markou A, Kosten TR, Koob GF (1998) Neurobiological similarities in depression and drug-dependence. *Neuropsychopharmacol* 18: 135–174

72   Hökfelt T, Pernow B, Wahren J (2001) Substance P – a pioneer amongst neuropeptides. *J Intern Med* 249: 27–40

73   Holsboer F (2001) Prospects for antidepressant drug discovery. *Biol Psychol* 57: Sp iss 47–65

74   Makino S, Hashimoto K, Gold PW (2002) Multiple feedback mechanisms activating corticotropin-releasing hormone system in the brain during stress. *Pharmacol Biochem Behav* 73: 147–158

75   Heuser I (1998) Anika-Monika-Prize paper. The hypothalamic-pituitary-adrenal system in depression. *Pharmacopsych* 31: 10–13

76   Mitchell AJ (1998) The role of corticotropin-releasing factor in depressive illness – a critical review. *Neurosci Biobehav Rev* 22: 635–651

77   Scott LV, Dinan TG (2002) Vasopressin as a target for antidepressant development: an assessment of the available evidence. *J Affect Disorders* 72: 113–207

78   Steiger A, Holsboer F (1997) Neuropeptides and human sleep. *Sleep* 20: 1038–1052

79   Steiger A, Antonijevic IA, Bohlhalter S, Frieboes RM, Friess E, Murck H (1998) Effects of hormones on sleep. *Hormone Res* 49: 125–130

80   McDougle CJ, Barr LC, Goodman WK, Price LH (1999) Possible role of neuropeptides in: obsessive-compulsive disorder *Psychoneuroendocrinol* 24: 1–24

81   Rook GAW, Lightman SL, Heijnen CJ (2002) Can nerve damage disrupt neuroendocrine immune homeostasis? Leprosy as a case in point. *Trends Immunol* 23: 18–22

82   Naafs B (2000) Current views on reactions in leprosy. *Indian J Lepr* 72: 97–122

83   Tenbokum AMC, Hofland LJ, Vanhagen PM (2000) Somatostatin and somatostatin receptors in the immune system – A review. *Eur Cytokine Network* 11: 161–176

84   Lotti T, Bianchi B, Ghersetich I, Brazzini B, Hercogova J (2002) Can the brain inhibit inflammation generated in the skin? The lesson of alpha-melanocyte-stimulating hormone. *Intern J Dermatol* 41: 311–318

85   Huygen FJPM, Debruijn AGK, Klein J (2001) Neuroimmune alterations in the complex regional pain syndrome. *Eur J Pharmacol* 429: Sp Iss: 101–113

86   Vrinten DH, Adan RA, Groen GJ, Gispen WH (2001) Chronic blockade of melanocortin receptors alleviates allodynia in rats with neuropathic pain. *Anesthesia and Analgesia* 93: 1572–1577

87   Catania A, Airaghi L, Garofalo L, Cutuli M, Lipton JM (1998) The neuropeptide alpha-MSH in HIV- infection and other disorders in humans. *Ann NY Acad Sci* 840: 848–85

88   Strand FL, Rose KJ, King JA, Segarra AC, Zuccarelli LA (1989) ACTH modulation of nerve development and regeneration. *Prog Neurobiol* 33: 45–85

89   Strand FL, Lee SJ, Lee TS, Zuccarelli LA, Antonawich FJ, Kume J, Williams, KA (1993) Non-corticotropic ACTH peptides modulate nerve development and regeneration. *Rev Neurosci* 4: 321–364

90   Strand FL, Williams KA, Alves SE, Antonawich FJ, Lee TS, Lee SJ, Kume J, Zuccarelli, LA (1996) Melanocortins as factors in somatic neuromuscular growth and regrowth. In: C Bell (ed): *Chemical factors in neural growth, degeneration and repair.* Elsevier, Amsterdam, 311–337

91   Azmitia, EC, de Kloet E (1987) ACTH neuropeptide stimulation of serotinergic maturation in tissue culture: modulation by hippocampal cells. *Prog Brain Res* 72: 311–318

92   Lee S J, Lee T S, Strand F L (1991) Local control of neurite outgrowth of dorsal root ganglia and spinal cord neurons by ACTH analog Org 2766, BIM 22015 and NGF. *Soc Neurosci Abstract* 598: 12

93   Van der Neut R E, Hol M, Gispen W H, Bar P R (1992) Stimulation by melanocortins of neurite outgrowth from spinal and sensory neurons *in vitro*. *Peptides* 13: 1109–1115

94 Beckwith BE, Sandman CA, Hothersall D, Kastin AJ (1977) The influence of neonatal injections of α-MSH on learning, memory and attention in rats. *Physiol Behav* 18: 63–71

95 Rose KJ, Frischer RE, King JA, Strand FL (1988) Neonatal neuromuscular parameters vary in susceptibility to ACTH/MSH 4-10 administration. *Peptides* 9: 151–156

96 Segarra AC, Luine VN, Strand FL (1991) Sexual behavior of male rats is differentially affected by timing of perinatal ACTH administration. *Physiol Behav* 50: 689–697

97 Alves SE, Akbari HM, Azmitia EC, Strand FL (1993) Neonatal ACTH and corticosterone alter hypothalamic monoamine innervation and reproductive parameters in the female rat. *Peptides* 14: 379–384

98 Flohr H, Lüneburg U (1989) Influence of melanocortin fragments on vestibular compensation. In: M Lacour, M Toupet, P Denise et al (eds): *Vestibular compensation: facts, theories and clinical perspectives*. Elsevier, Paris, 161–174

99 Nyakas C, Veldhuis HD, DeWied D (1985) Beneficial effects of chronic treatment with ORG 2766 and α-MSH on impaired reversal learning of rats with bilateral lesions of their parafascicular area. *Brain Res Bull* 15: 257–265

100 Hannigan J, Isaacson R (1985) The effects of ORG 2766 on the performance of sham, neocortical and hippocampal-lesioned rats in a food search task. *Pharmacol Biochem Behav* 23: 1019–1027

101 Attella M J, Hoffman S W, Pilotte M P, Stein DG (1992) Effects of BIM 22015, an analog of ACTH 4-10, on functional recovery after frontal cortex injury. *Behav Neural Biol* 57: 157–166

102 Antonawich FJ, Azmitia EC, Strand FL (1993) Rapid neurotrophic actions of an ACTH/MSH (4-9) analog after nigrostriatal 6-OHDA lesioning. *Peptides* 14: 1317–1324

103 Wolterink G, Van Zanten E, Kamsteeg K, Radhakishun FS, Vanree JM (1990) Functional recovery after destruction of dopamine systems in the nucleus accumbens of rats. II. Facilitation by the ACTH- (4-9) analog ORG 2766. *Brain Res* 507: 101–108

104 Antonawich FJ, Azmitia EC, Kramer HK, Strand FL (1994) Specificity versus redundancy of melanocortins in nerve regeneration. *Ann NY Acad Sci* 739: 60–73

105 Van der Hoop RG, Vecht CJ, Van der Burg MEL, Elderson A, Boogerd W, Heimans JJ, Vries EP, Van Houwelingen JC, Jennekens FGI, Gispen WH et al (1990) Prevention of cisplatin neurotoxicity with ACTH (4-10) analogue in patients with ovarian cancer. *N Engl J Med* 322: 89–84

106 Bär PRD, Schrama LH, Gispen WH (1990) Neurotropic effects of ACTH/MSH-like peptides in the peripheral nervous system. In: D DeWied (ed): *Neuropeptides: basics and perspectives*. Elsevier, Amsterdam, 175–211

107 Strand FL, Stoboy H, Friedebold G, Krivoy W, Heyck H, Vanriezen H (1977) Changes in muscle action potentials of patients with diseases of motor units following the infusion of a peptide fragment of ACTH. *Drug Res* 27: 681–683

Progress in Drug Research, Vol. 61
(L. Prokai and K. Prokai-Tatrai, Eds.)
©2003 Birkhäuser Verlag, Basel (Switzerland)

Structural and functional aspects of the blood-brain barrier

By David J. Begley[1] and Milton W. Brightman[2]

[1]Centre for Neuroscience Research
Kings College London
Hodgkin Building
Guy's Campus, London SE1 1UL
UK
<david.begley@kcl.ac.uk>

[2]National Institutes of Health
Building 36, Room 2A-21
Bethesda, MD 20892-4062
USA
<brightmm@ninds.nih.gov>

David Begley

received his BSc (1967) and PhD (1971) Degrees from the University of Nottingham, UK. At present he is Senior Lecturer in Physiology at Kings College London and is Co-Director (With Prof N.J. Abbott) of the Blood-Brain Barrier Research Group within the Centre for Neuroscience Research at Kings College. He is author of more than fifty key papers on blood-brain barrier function and drug delivery to the CNS and has contributed fourteen chapters to edited volumes. He has also edited three volumes on the topics of blood-brain barrier function and CNS drug delivery and has organised a number of International Research Symposia. He is organiser of the UK Physiological Society Special Interest Group on the Blood-Brain Barrier. His interests encompass the general function of the blood-brain and CSF barriers, transport systems in the CNS and its barriers and drug targeting to the brain and spinal cord. He was organiser and Chairman of the Gordon Conference on "Barriers of the CNS" held in New Hampshire in 2002. He lectures frequently in Europe and the USA on the blood-brain barrier and receives major research support as funded consortia with the pharmaceutical industry on an international basis and also The Wellcome Trust and European Union projects involving Germany, France, The Netherlands and The Russian Federation. He was the Friedrich Mertz Stiftungsgastprofessor, Johann Wolfgang Goethe-Universität, Frankfurt, academic year 1997–1998 during the tenure of which he organised the international symposium "The Blood-Brain Barrier and Drug Delivery to the CNS" at that University and he is currently Honorary Docent and visiting Professor, Institut für Pharmazeutische Technologie, Johann Wolfgang Goethe-Universität, Frankfurt. He is a member of the Editorial Board of The Journal of Drug Targeting. *His interests outside of science include yachting, gardening and fine food and wine.*

Milton Brightman

received his Ph.D. in the Department of Anatomy, Yale University in 1953, joined the Laboratory of Neuroanatomical Sciences at the National Institutes of Health the following year and became Head of the Section on Neurocytology in 1969. His publications of over 100 papers include the reciprocal synapses of the olfactory bulb, the structural basis of the blood-brain and blood-CSF barriers, and beyond the barriers, the accessibility of the interstitial clefts of the central nervous system to probe molecules introduced into the CSF. The astroglial walls of the clefts and the glycoprotein content of the clefts are current interests. He served as associate editor on a number of Journals and as the American Editor of the Journal of Neurocytology *from 1992 to 2002. Dr. Brightman became Scientist Emeritus in 1996 and continues to edit and write.*

Contents

Key words

ABC transporters, blood-brain barrier, blood-CSF barrier, cerebrospinal fluid, choroid plexus, endocytosis, endothelium, epithelium, exocytosis, interstitial fluid, solute transporters, tight junctions, transport.

Glossary of abbreviations

4F2hc, unbiquitous cell surface antigen (CD98) heavy chain; A/ATA, amino acid transporter (preferring small neutral amino acids and an N-methyl group); ABC, ATP binding cassette; ACE, angiotensin converting enzyme; AMT, adsorptive-mediated transcytosis; ASC, neutral amino acid transporter (preferring alanine, serine and cysteine); ATP, adenosine triphosphate;

β, β-amino acid transporter; BBB, blood-brain barrier; BCRP, breast cancer-resistance protein; BCSFB, blood-cerebrospinal fluid barrier; bFGF, basic fibroblastic growth factor; $B^{o,+}$, neutral/cationic amino acid transporter; BUI, brain uptake index; cAMP, cyclic adenosine monophosphate; CAT1, cationic amino acid transporter; cGMP, cyclic guanine mono phosphate; cib, concentrative nucleoside transporter (broad spectrum of substrates); cif, concentrative nucleoside transporter (formycin-b preferring); cit, concentrative nucleoside transporter (thymidine preferring); CNS, central nervous system; CNT, concentrative nucleoside transporter; CSF, cerebrospinal fluid; DNA, desoxyribose nucleic acid; EAE, experimental allergic encephalitis; ECF, extracellular fluid; ei, equilibrative nucleoside transporter (insensitive to NBMPR); ENT, equilibrative nucleoside transporter; es, equilibrative nucleoside transporter (sensitive to NBMPR); F-actin, filamentous actin; GLUT, glucose transporter; gp120, glycoprotein 120; GSH, glutathione; HIV-1, human immunodeficiency virus-1; HRP, horseradish peroxidase; ICAM-1, intercellular adhesion molecule-1; IFN-γ, gamma interferon; IL-1β; interleukin 1β; IL-3, interleukin-3; ISF, interstitial fluid; JAM, junction-associated molecule; kDa, kilo-Daltons; K_m, Michaelis-Menten constant; L, large neutral amio acid transporter; LAT, large neutral amino acid transporter; LNAA, large neutral amino acid; MCT, monocarboxylic acid transporters; MDR, multidrug resistance; mM, millimolar; mRNA, messenger RNA; MRP, multidrug resistance-associated protein; MXR1, mitoxantrone-resistance protein = BCRP; NBMPR, nitrobenzylmercatopurine riboside; NGF, nerve growth factor; OAT, organic anion transporter; Oatp, organic anion transporting polypeptide; OCT, organic cation transporter; OCTN, organic caion transporter-novel; PEPT2, peptide transporter protein-2; Pgp, P-glycoprotein; RMT, receptor-mediated transcytosis; RNA, ribose nucleic acid; SGLT1, sodium-dependent glucose transporter-1; SVCT2, nucleobase transporter-2; system β, amino acid transporter (preferring β amino acids); system L, large neutral amino acid tansporter; TER, transendothelial/epithelial resistance; TGF-β, transforming growth factor β; TNFα, tumour necrosis factor α; UDP, uridine diphosphateglucuronosyltransferase; VCAM-1, vascular cell adhesion molecule; VLA-4, very late antigen-4; WGA, wheat-germ agglutinin; X^-_{AG}, amino acid transporter (preferring glutamic acid and aspartic acid); y^+, cationic amino acid transporter; ZO, zona occludentes; ZO1/2/3, zona occudentes protein 1/2/3.

1   Introduction

All organisms with a complex nervous system have a well-developed blood-central nervous system barrier. In the vertebrates the central nervous system (CNS) lies behind the protective blood-brain barrier (BBB) and blood-cerebrospinal fluid barrier (BCSFB). In all mammals the BBB is formed at the level of the cerebral capillary endothelial cells and the BCSFB by the choroid plexus epithelium.

The physical barrier to solute movement between the cerebral endothelial cells is formed by transmembrane proteins creating tight junctional complexes between the cells which effectively abolishes any aqueous paracellular diffusional pathways between the blood and brain extracellular fluid [1]. In a complementary way a diffusional barrier is created between the apical epithelial cells of the choroid plexus by similar junctional complexes creating a comparable diffusional barrier between the extracellular fluid of the choroid plexus, which is in free diffusional communication with the blood plasma, and the cerebrospinal fluid (CSF) [2]. An essential function of the BBB and the BCSFB is to create a separate extracellular fluid (ECF) compartment for the CNS distinct from the general somatic ECF, which can be regulated with precision in terms of its solute composition. The CNS relies almost entirely for its function on accurate synaptic transmission, combined with summation and inhibition, all of which are chemically regulated. Unless the synapse can operate against a very stable fluid background environment, the complex integrative and control functions of the CNS become impossible to sustain. A simple example of the BBB at work is that of neuromuscular blockade. This is pharmcologically relatively easy to achieve with a number of injected agents acting both pre- or post-synaptically at the neuromuscular junction. However in most cases the central cholinergic and other synapses are unaffected.

The creation of the BBB and BCSFB therefore effectively seals CNS from most polar blood-borne solutes. Thus, in order for the CNS to receive a sufficient supply of vital polar metabolites and other solutes necessary for CNS function, the cerebral vasculature must contain specific transport mechanisms to deliver these solutes across the barriers [3].

A significant number of lipophilic solutes are able to diffuse across the brain barrier by direct permeation through the lipoidal cell membranes. In general, the more lipid soluble a molecule is the greater is its potential for passively partitioning into brain tissue [4]. However, many lipid soluble solutes do not penetrate into brain to an extent that might be anticipated from their lipid solubility. These solutes, plus some of their metabolites, are actively removed from the cerebral compartment by efflux transporters in the blood-brain and blood-CSF barriers [5–7].

A further general function of the blood-brain barriers is that of neuroprotection. During a lifetime the CNS will be exposed to a large number of physiological metabolites and acquired xenobiotics, many of which are potentially neurotoxic. In a tissue such as the CNS, where in the adult neuronal cell division is virtually absent, any acceleration in cell death will cause

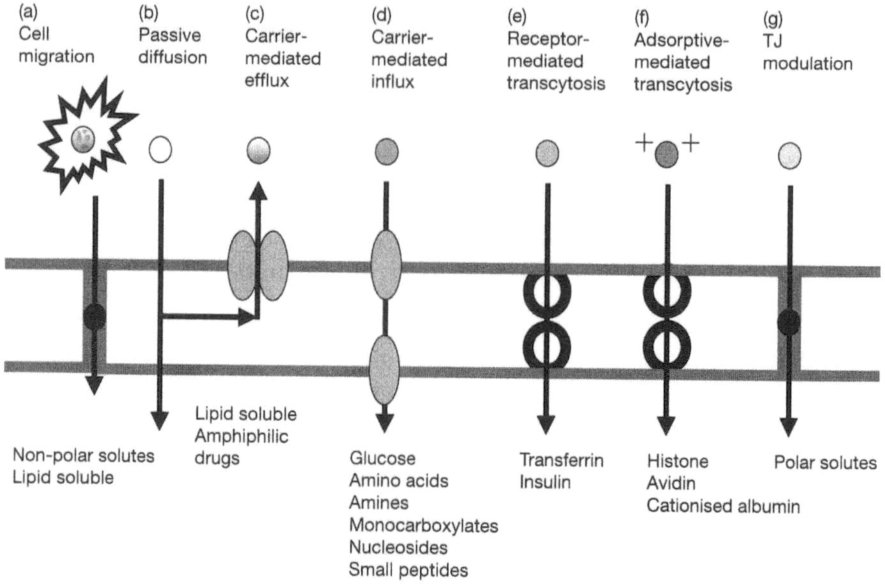

Figure 1.
Transport across the blood-brain barrier. (a) Leukocyte movement across the BBB; (b) Molecules may diffuse across the BBB by passive diffusion through the cell membranes and across the endothelial cells. In general the more lipid soluble a molecule the greater the CNS penetration; (c) Active carrier-mediated efflux transporters are able to expel a wide and varied range of molecules out of the BBB and reduce their CNS penetration to a level below that predicted by their lipid solubility; (d) Many necessary polar metabolites and nutrients must be carried into the CNS by transporters in the BBB; (e) Some macromolecules, proteins and peptides are transported by receptor-mediated transcytosis (RMT); (f) Other macromolecules, especially cationic, induce adsorptive-mediated transcytosis (AMT); (g) Tight junctions may be modulated to allow an increased movement of polar solutes through the aqueous paracellular pathway. These mechanisms and others are discussed in the text.

premature degenerative disease and pathology. The formation of a barrier to these potentially toxic substances, both in terms of the physical barrier and the metabolic and transport barriers, serves to protect the CNS.

There is normally very little obvious vesicular, transcytotic traffic across the CNS endothelium, but some solutes can be brought across by receptor-mediated transcytosis in small vesicles. Another means of solute entry into brain does not involve the CNS vasculature but rather an extracellular, peri-axonal pathway from peripheral tissues such as muscle and nasal mucosa.

An account of these barriers forms the basis of this review and is summarised in Figure 1.

2   Tight junctions

The locus of the endothelial or blood-brain [8], and the epithelial or the blood-CSF [9, 10], paracellular barriers is the tight junction. A popular method used to assess the integrity of tight junctions and thus permeability of an endothelial or epithelial monolayer *in vitro* is to measure the trans-endothelial (or epithelial) electrical resistance (TER). In most current *in vitro* preparations, brain endothelial cells are co-cultured with a layer of astrocytes to yield, most commonly, a TER of about 100–150 $\Omega \cdot cm^2$ as compared to the mean value of 1800 $\Omega \cdot cm^2$ for pial microvessels *in vivo* [11].

The endothelial and epithelial tight junction consists of three integral membrane proteins : claudins, occludin and a junctional adhesion molecule (JAM) which, together with several peripheral proteins, constitute the junctional complex. The appearance of the formed tight junction is most readily appreciated in replicas of the frozen and fractured cell membranes at the junction (reviewed in [12]). Within the cleaved lipid bilayer of endothelial cell membrane from brain, the junction appears as a series of parallel, interconnected strands or fibrils (Fig. 2). The formation of the strands depends on claudin-5 (23 kDa) [13]. The tight junctions of choroid plexus epithelium differ from endothelium in that they also contain claudins –1 and 11 [2]. Occludin (65 kDa) [14], is not required for strand formation [15] but, in some ill-defined way, affects junctional permeability. JAM (36–41 kDa), a member of the immunoglobulin superfamily, is involved in monocyte transmigration. In brain, the indirect evidence for this function has been obtained in mice with meningitis induced by the intraventricular infusion of the inflammatory cytokines TNF-α and IL-1β. The prior intravenous injection of a monoclonal antibody to JAM, attenuates the migration of neutrophils and monocytes into the CSF and brain parenchyma of these mice [16].

A number of peripheral proteins, regarded as scaffolding or linkers for the integral molecules of the tight junction in endothelia and epithelia, form a cytoplasmic plaque adjacent to the junctional cell membrane. The plaque proteins are part of the tight junction complex and include the guanylate kinases: ZO-1 (220 kDa), ZO-2 (160 kDa) and ZO-3 (130 kDa). Another plaque protein is cingulin (140 kDa), which is myosin-like and binds to the ZO complex and to F-actin of the cytoskeleton [17].

Except where there are fibrillar discontinuities (Fig. 2), which imply increased permeability, the number, length and complexity of the strands

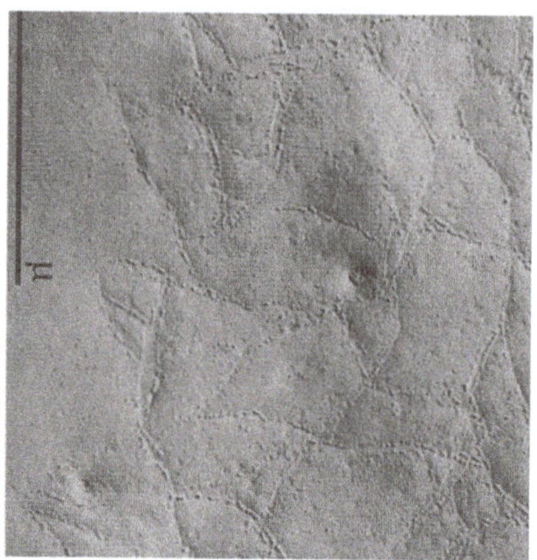

a   b

Figure 2.
Carbon and platinum replicas of frozen-fractured confluent layers of bovine brain endothelial cells *in vitro*. Tight junctions appear as granular strands or fibrils.
(a) In developing tight junctions (left figure), the strands have loose ends (arrows), indicative of a permeable junction, which also has the immature attribute of an intercalated gap junction (GJ), consisting of larger particles. (b) This mature tight junction appears as a network of interconnected strands or fibrils. The line marked μ refers to 1 μm [54].

and the ZO-1 linker phosphoprotein cannot be consistently correlated with the permeability of the tight junction. Epithelia with a high TER and thus relatively impermeable to solutes, can be rendered leaky *in vitro*. Their TER falls, yet the arrangement and pattern of the junctional fibrils are identical to those cell layers with a high TER [18]. In two different strains of Mardin-Darby canine kidney cells, one having a fifty to seventy fold higher TER than the other, the strands are very similar in number, total length and complexity in both cell strains, as is the amount of ZO-1 [19].

Although the tight junction's fibrillar substructure does not reflect its permeability, its state of phosphorylation does. Phosphorylation of the proteins constituting tight junctions enhances their permeability, as assessed *in vitro*, by the passive flux of extracellular probes such as albumin (67 kDa) or inulin (5 kDa) and by the TER across the cell monolayer. Tyrosine phosphorylation can be augmented in the adhesion plaque protein, paxillin, which links integrins to the cytoskeleton (Fig. 2). This phosphorylation, e.g. in isolated coronary venules, leads to an increased passive permeability of the endothelium to albumin [20]. In addition, enhanced tyrosine phosphorylation in brain endothelial cells, decreases their TER [21].

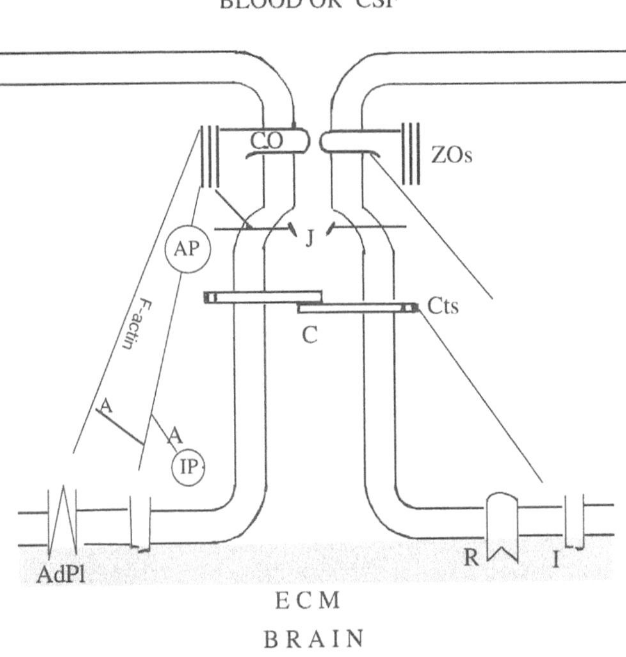

Figure 3.
Tight junction and adherens junction between endothelial cells and between epithelial cells. Tight junctions consist of integral (intramembranous) membrane proteins, CO (Claudins and Occludin) and the junction adhesion molecule or JAM, labeled here as J. Three peripheral proteins, ZO-1, 2, and 3 of the tight junction (or Zonula Occludens) are closely associated with the integral proteins.
A, α-actinin protein cross-links F-actin to other F-actin filaments and to other plasma membrane components, such as the adhesion molecules of integrins and adhesion plaques; Ad Pl, adhesion plaques situated at the basal surface of the two cells face the ECM; ECM, extracellular matrix, includes basal lamina lying immediately beneath the cells; AP, accessory proteins of the tight junction, e.g. cingulin; C, cadherins of the adherens junction; Cts, catenins are linker proteins of the adherens junction; ECM, extracellular matrix includes the basal lamina, depicted as the shaded band; I, integrins; IP, integrin associated proteins; J, junction associated molecule (JAM); OC, occludin and claudins; R, receptor, e.g. for growth factor; ZOs, linker proteins of Zonula Occludens (tight junction).

Extracellular signals can influence the permeability of the tight junction by a common structural path, the cytoskeleton, which connects adhesion molecules and integrins on the cell membrane to the molecules comprising the junctional complex (Fig. 3). The peripheral protein ZO-1 provides a link between its integral membrane protein, occludin, by way of F-actin [22], a component of the cytoskeleton (Fig. 3). Cingulin, a myosin-like protein binds to the ZO complex and to actin [17], thereby further connecting the tight junction to the cytoskeleton. The tight junction is, therefore, connected to the cytoskeleton by way of both its integral and its peripheral subunits. The molecular anatomy of the endothelial junctional complex of the blood-brain barrier has recently been reviewed [23].

3 Cytokines

The blood-brain barrier is reversibly disrupted during inflammation by cytokines, which are humoral substances secreted by leukocytes from peripheral blood and by cells of the CNS. Astrocytes, co-cultured with brain endothelial cells, secrete cytokines, including basic fibroblastic growth factor (bFGF) and when stimulated, secrete a combination of pro-inflammatory cytokines: TNF-α, transforming growth factor-beta (TGF-β), and interleukins, e.g. IL-3 [24].

Cytokines open endothelial tight junctions by acting directly on the integral molecules of the junction and, by also acting on their ZO linkers, they affect cytoskeletal actin. An alternate route taken by signals, initiated at the basolateral cell membrane where focal adhesions, integrins and certain receptors reside, is by way of actin to the junction complex (Fig. 2). When TNF-α is added to the luminal face of a brain endothelial monolayer *in vitro*, there is a concomitant increase in permeability and a reorganization of F-actin into stress fiber bundles. The significance of the actin reorganization may be a change in cell shape which could affect the configuration of the tight junction [25].

A mechanism by which the permeability of an epithelium or endothelium is reversibly elevated appears to be the temporary translocation of its junctional proteins away from the site of the tight junctions. The tight junction-associated protein (JAM) is redistributed from the junction region of the cell membrane toward the apical surface, following a combined treatment with TNF-α and IFN-γ (gamma interferon). The result is an increase in leukocyte passage across endothelium [26].

4 Cell entry

The barrier endothelium that restricts polar solutes from entering normal brain also restrains circulating immunocytes. Activated cells, nevertheless, can penetrate the brain. T-lymphocytes, stimulated non-specifically, *in vitro*, with the mitogenic lectin, concanavalin-A and re-injected into blood, enter CNS randomly then leave. T cells not only enter the CNS but are retained within it only after they are presented with and recognize brain-specific antigens [27].

Passage across endothelium must be preceded by temporary adhesion of the lymphocytes to the endothelium by way of cell adhesion molecules expressed on both immunocytes and endothelium. The more than 20 adhesion molecules so far identified include members of the immunoglobulin family, e.g. vascular cell adhesion molecule 1 (VCAM-1) and intercellular adhesion molecule 1 (I CAM-1) on human brain endothelium [28] as well as in rodents. The ligand for endothelial VCAM-1 is very late antigen-4 (VLA 4), an integrin receptor expressed, e.g. on monocytes and lymphocytes. The potentially therapeutic, immunological blocking of VLA-4 reduces the number of lymphocytes binding to inflamed CNS in animals with experimental allergic encephalitis (EAE). A weakness of this interpretation is that antibodies to the adhesion factors may be given long after intravascularly infused encephalitogenic T cells have penetrated the CNS, and so it would appear that their effect may not be on adhesion alone [27]. Lymphocytes that adhere to the luminal face of endothelium are now ready for transendothelial migration which can be enhanced by proinflammatory cytokines.

Immunocytes migrate across CNS endothelium both paracellularly and transcellularly. Paracellular migration of cells through the tight junctions has been inferred from blocking with antibody against its associated junctional adhesion molecule (JAM). The result is a decreased accumulation of cytokine-activated leucocytes in CSF and parenchyma [16] as discussed above. A paracellular but non-junctional route for activated neutrophils is through extracellular "corners" between three contiguous human umbilical endothelial cells grown in astrocyte conditioned medium containing cytokines. The transendothelial resistance remains unchanged during cell passage, so the path may not be junctional [29]. CNS endothelium was not used in these thoroughly done experiments, which, therefore, can only be extrapolated to brain endothelium. A transendothelial passage of neutrophils across pial vessels *in situ* was brought about by stimulating these vessels which emit arm-like evaginations that embrace the cells and appear to draw them through the endothelial cells rather than between them [30].

The *in vivo* site of leukocyte migration during inflammation is determined by a subtle difference in the composition of the laminin component of the perivascular basement membrane. In mice with EAE, the laminin bordering the perivascular space of post-capillary venules contains only the α-4 chain. It is here where activated, circulating T lymphocytes traverse the endothelium and its basal lamina, to enter the perivascular space. The mRNA for the α-4 chain is enhanced by TNF-α and IL-1 [31]. Once in the perivascular space, the cells continue their penetration into the CNS parenchyma by secreting matrix metalloproteases. These enzymes degrade the extracellular matrix thereby enabling the cells to cross the glia limitans to enter the parenchyma.

5 Vesicular transport

Although certain ions and solutes are selectively transferred across CNS endothelium by special carriers (see below), some solutes and viruses are transported across cells by vesicular transcytosis. Transcytosis begins as a microinvagination of the cell membrane to form a pit or caveola (Fig. 4). The pit pinches off as a free vesicle that migrates across the cell to the opposite cell membrane to which it fuses and releases its contents into the periendothelial basal lamina. What little solute is incorporated by fluid-phase endocytosis in brain endothelium is usually destined for enzymatic degradation in cytoplasmic organelles such as lysosomes.

5.1 Adsorptive transcytosis

The luminal portion of brain endothelial cell membrane is rich in galactosylated glycoconjugates which bind lectin, such as wheat germ agglutinin (WGA). A conjugate of this lectin and horseradish peroxidase (HRP) are adsorbed to the glycoconjugates and the entire complex is endocytosed. Such adsorptive endocytosis and efflux of lectins is more extensive but much slower than fluid-phase endocytosis of HRP and requires energy [32]. Non-permeating solute conjugated to lectins are not only internalized but may be transcytosed across brain endothelium [33]. Another mechanism for adsoptive endocytosis is the salt-linkage between the negatively charged sialic acid [34] and phosphate groups on the endothelial's luminal face and positively charged solutes. Cationization of proteins such as albumin increases the rate and amount at which it is transcytosed [35]. However, the cationization of albumin is also necessary because CNS endothelium lacks the albumin receptor [36] that the endothelia of other organs have [37].

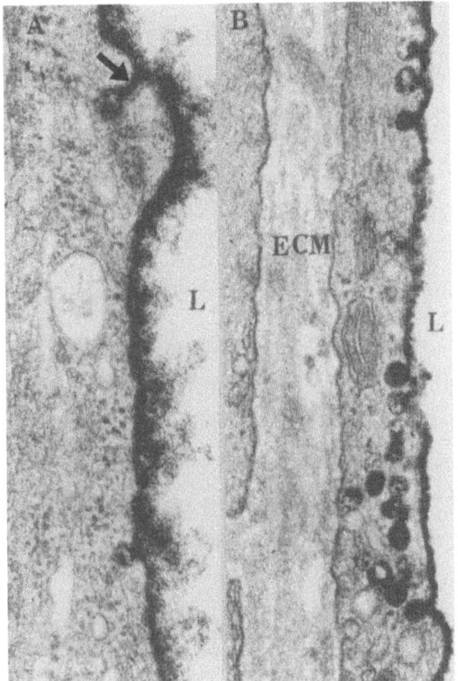

Figure 4.
Caveolae in endothelium of an arteriole in rat brain.
(A) Horesradish peroxidase (HRP), a fluid-phase marker, when injected intravenously, reaches brain vessels where it enters the long neck (arrow) of a caveolar pit confluent with the vessel lumen (L).
(B) Unlike capillary endothelium, this arteriolar endothelium is replete with caveolar pits. Many of the pits contain HRP. Without serial sections, pits cannot be differentiated from vesicles lying free within the cytoplasm. No detectable HRP was transcytosed at this time, as indicated by the absence of free reaction product in perivascular extracellular matrix (ECM) [55].

5.2 Receptor mediated transcytosis

A specific transfer of solutes across CNS endothelium is mediated by receptor transcytosis; a ligand binds to its receptor on the endothelial cell membrane which forms a vesicle that is transcytosed. Transferrin, an iron-transporting metalloprotein, has receptors on the surface of blood vessels throughout the brain. Advantage has been taken of the receptors' ubiquitous distribution by conjugating a ligand to an antibody to the receptor in order to bring it into brain [38]. The efficiency of this transfer is low in the mature rat brain, where "the transcytosis of transferrin into the brain interstitium is only a minor pathway" [39]. Nevertheless, in rats with induced focal ischemia, the intravenous infusion of brain-derived neurotrophic factor, conjugated to the transferrin receptor antibody, reduces the volume of ischemic injury by about 68% to 70% [40]. The reparative stages following ischemia might mimic normal development by inducing a transitory upregulation of transferrin receptors.

5.3   Protein entry into brain *via* cranial and spinal nerves

In the CNS, endocytosis is performed by endothelium and certain neurons in the brain stem, the axons of which project to the periphery or to small regions of the brain that lack a blood-brain barrier. These regions include the neural lobe (Fig. 5), median eminence, area postrema and choroid plexus.

The brain stem nuclei become labeled with circulating horseradish peroxidase (HRP) (Fig. 6), if its plasma concentration is sufficiently high and maintained for about 12 to 24 hours in mice [41]. But the HRP does not cross the barrier endothelium of the CNS. Instead, the HRP is endocytosed directly by axonal and dendritic terminals of cranial and spinal nerves that become bathed by HRP exuding across the permeable vessels of peripheral tissue, such as muscle, innervated by the axons. Similarly, the brainstem neurosecretory axons, arising from the supraoptic and paraventricular nuclei, terminate upon fenestrated, permeable vessels of the pituitary gland's neural lobe from which they endocytose the circulating HRP and transport it back to their cell bodies [41] (Figs. 5 and 6).

5.4   Viral passage

Viruses can penetrate the blood-brain barrier to enter brain parenchyma in several ways: e.g. paracellularly, vesicularly and as leukocyte cargo. This discussion is limited to the paracellular and vesicular entry. The paracellular route for non-neurotropic viruses becomes available during inflammation. When the inflammatory cytokine, TNF-α is added to the medium of brain endothelium *in vitro*, HIV-1 (human immuno-defficiency virus-1), detected by its RNA, is able to enter the endothelial cells. The paracellular pathway, confirmed by the concurrent passage of the extracellular markers, inulin (5 kDa) and dextran [42], is not available to this virus across normal endothelium.

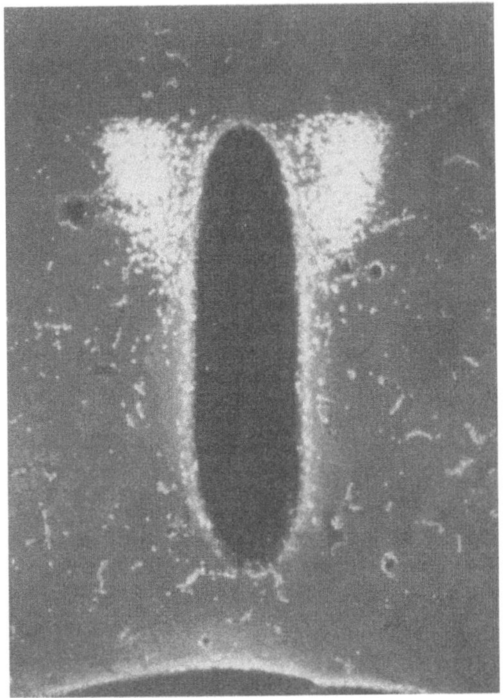

Figure 5.
Retrograde labeling of brainstem nucleus from blood. The neurosecretory paraventricular nucleus of the hypothalamus in a mouse has endocytosed circulating HRP from the perivascular spaces of the pituitary gland's neural lobe. This HRP had exuded from the permeable, fenestrated vessels of the neural lobe and was transported by retrograde axoplasmic flow to the neuronal cell bodies. Within the cell bodies, the HRP is deposited in lysosomes, thereby visibly labeling the neurons before being hydrolysed in the lysosomes. About 50 mg of HRP had been intravenously infused in mature mice and permitted to circulate for 12–24 hours before fixation [41, 56]. Dark-field illumination.

Non-neurotropic viruses, such as HIV-1, can be transcytosed across unperturbed endothelium *in vivo*. Virion binding to the endothelial cell membrane as a first step is critical. The HIV-1 virion's envelope glycoprotein gp120 binds to the cell membrane before the endocytotic vesicle forms. HIV-1 without gp120 does not bind and is not transcytosed. The paracellular route is inaccessible in normal animals, as manifested by the failure of albumin, co-infused with virus, to enter brain [43].

Another vesicular route available to circulating virus is axonic rather than endothelial. The fluid-phase endocytosis of HRP from the blood by axon terminals and its retrograde axoplasmic flow to brainstem nuclei [41] has been

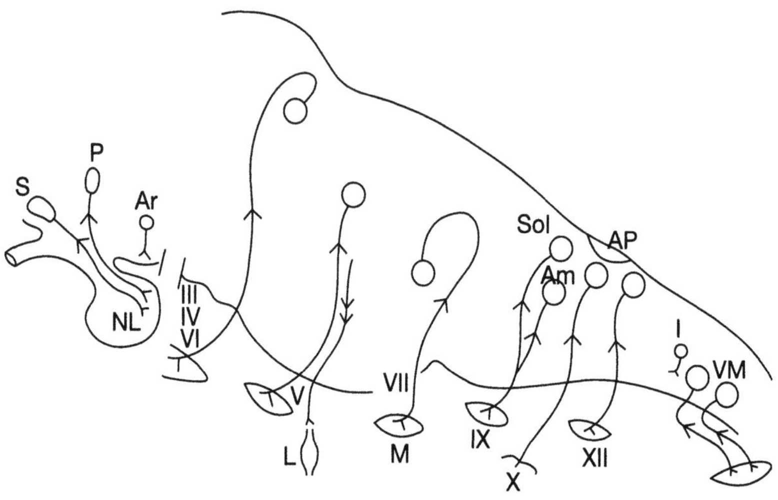

Figure 6.
Entry of circulating protein into brainstem nuclei. Circulating HRP leaves the permeable blood vessels of peripheral tissue, e.g. muscle and the brain's circumventricular organs, e.g. neural lobe (NL) and median eminence. The arcuate nucleus is labeled from vessels in the median eminence and the neurosecretory nuclei: S and P from pituitary blood. Note that protein enters the trigeminal nerve (V) motor nucleus via its axons but protein from within the brain leaves the brain by flowing extra-axonally (double arrow) along this nerve and cranial nerves I, II and VIII.
It is emphasized that intra-axonal entry into brainstem is confined to the nuclei and does not leave neurons to spill into the interstitial clefts. Based on [41], modified from [56].
Am, nucleus ambiguus; AP, area postrema; Ar, arcuate nucleus; I, intermediolateral column of spinal cord; L, deep cervical lymph node; M, muscle; NL, neural lobe of the pituitary gland; P, paraventricular nucleus of hypothalamus; S, supraoptic nucleus of hypothalamus; Sol, nucleus of the tractus solitarius; VM, ventral motor horn cells of the spinal cord.

suggested as a means by which circulating, non-neurotropic virus can enter brain during a prolonged viremia [44]. The HRP analogy is plausible for viral entry because, in the early stages of HIV-1 infection, a circulating level of 10^7 viral RNA copies/ml of blood is attained [42]. The analogy also implies that the viremia would have to be sustained for 12 to 24 hours.

It is emphasized that the HRP is confined to the cranial nerve nuclei within the brain stem. However, when conjugated to the lectin, the HRP is able to move trans-synaptically to, at least, second-order neurons [45]. Neurotropic viruses bind selectively to receptors on neurites and can, without the aid of lectin, also migrate trans-synaptically. The incorporation and trans-synaptic passage may be so specific that pseudorabies has been used to trace functionally discrete neural circuits [46].

5.5   Cranial nerve route into and out of brain

The blood-brain barrier may be non-invasively bypassed along an intracellular and an extracellular route involving the olfactory nerve. This route does not include the CNS endothelium. When HRP is placed in the nasal mucosa some of it is endocytosed by axon terminals and retrogradely transported within the axon to the cell body in the olfactory bulb [45]. HRP, conjugated to a lectin, is adsorptively endocytosed, conveyed intra-axonally to its cell body, then trans-synaptically to other neurons within the olfactory bulb [45]. This intra-axonal route is a minor one for most solutes.

A major, concurrent, uptake of the solute by the same cranial nerve is extra-axonal. Solutes may diffuse extracellularly from the nasal mucosa into the peri-axonal sheaths of olfactory neurites. Extracellular passage is more rapid than the intracelluar one. The 26.5 kDa dimer of nerve growth factor (NGF), e.g. requires only 15 minutes to flow from nasal mucosa to the olfactory bulb in rats [47]; HRP (40 kDa), requires 45 to 90 minutes in rodents and squirrel monkeys. The slower intra-axonal route taken by lectin-conjugated HRP, requires about 6 hours to reach neurons in the olfactory bulb [45]. The fraction of NGF reaching the brain may only be picomolar in amount but is sufficient to enhance neural regrowth [47]. The size limit of solute that can be transferred in this way is about 20 kDa [48] but apparently includes the NGF dimer. It is not clear why there should be a size constraint for an extracellular route. Another limitation of the nasal route is the nasal mucosa's aminopeptidase which could degrade peptides [49], probably before they can migrate very far along the nerves.

The olfactory peri-axonal spaces are confluent with the sub-arachnoid space and thence the perivascular spaces and interstitial clefts throughout the brain [50]. The same route into brain is also available to the amoeba, *Naegleria fowleri*, which once in the nasal submucosa, can migrate peri-axonally, into the olfactory bulb within 24 hours in mice, to produce a necrotizing meningoencephalitis [51].

This extracellular pathway has immunological implications. The peri-axonal spaces that lead into brain also lead out of it along the sheaths of four cranial nerves: olfactory, optic, trigeminal and auditory, as well as spinal nerves (Fig. 5). Serum albumin, injected into the cerebral parenchyma or CSF, is propelled by bulk flow of interstitial fluid through the pathway described. About 14 to 47% of the albumin can be recovered from cervical lymphatics

[52]. Substances which could act as antigens, originating in the brain, could take this pathway which is confluent with extra-cerebral tissue spaces that are, in turn, confluent with lymphatics. The antigen can then reach regional lymph nodes to initiate an immune response [50, 52].

A view of how the blood-brain barrier affects the integrity of brain cells must take into account any consequences of the barrier's disruption. In addition to disturbances in solute homeostasis of the interstitial fluid, a protracted, repeated disruption could lead to an appreciable entry of plasma proteins into that fluid. This condition is approximated, *in vitro*, by adding albumin to a culture of astrocytes. The astrocytes respond by proliferating [53]. The *in vivo* response would be formation of an astroglial scar. It is suggested, therefore, that the absence of the albumin receptor on cerebral endothelium [36, 37] is yet another protective attribute of the CNS vasculature.

6   Fluid secretion at the blood-brain barriers

Fluid is secreted by both the cerebrovascular endothelium (BBB) and the epithelium of the choroid plexus (BCSFB). The fluid formed by the endothelial cells of the BBB contributes to the interstitial fluid of the brain (ISF) and that by the epithelium of the choroid plexus to the cerebrospinal fluid (CSF). The driving force for fluid formation in both cases is the sodium/postassium ATPase located, in the case of the BBB in the abluminal membrane of the endothelial cells [57] and in the apical membrane of the choroid plexus epithelium [58–60]. Sodium then enters the endothelium and the epithelium from blood down a concentration gradient by means of a Na^+/H^+ exchanger on the opposing membrane and is extruded into ISF or CSF by the Na^+/K^+ ATPase (Fig. 7). Chloride and bicarbonate can then follow sodium utilising either chloride channels, a sodium/chloride/potassium co-transporter or a chloride/bicarbonate exchanger (Fig. 7).

This secretion of Na^+, Cl^- and bicarbonate creates a potential osmotic gradient which drives water across the endothelium or epithelium. Water probably passes thorough the cells or may pass through the tight junctional complexes. Aquaporin 1 has been localised in the apical and basolateral membranes of the choroid epithelium [61, 62] and one study [63] has reported aquaporin 4 in the cell membranes of the BBB endothelial cells, although this has not been confirmed in other studies. Aquaporin 4 is certainly expressed in the perivascular end-feet and cell bodies of glial cells and in glial cells bor-

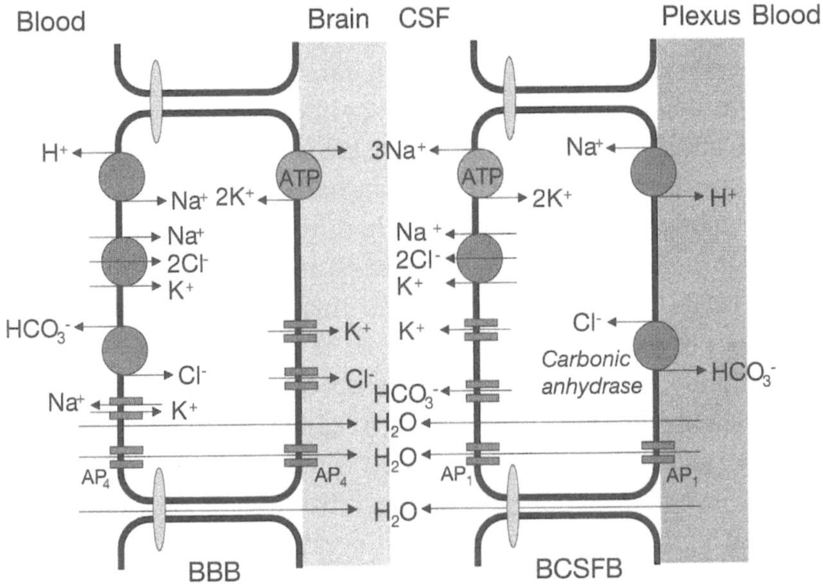

Figure 7.

Ion and water transport at the BBB and BCSFB. Fluid secretion is ultimately driven by the sodium/potassium-ATPase located in the abluminal membrane of the BBB and in the apical membrane of the choroid plexus. The epithelium of the choroid plexus is rich in carbonic anhydrase. Aquaporin 1 (AP$_1$) has been shown to be expressed in the choroid plexus. Only one study indicates aquaporin 4 (AP$_4$) expression in the BBB endothelium (see text).

dering the sub-arachnoid space and also the ependymal cells adjacent to the ventricles and in osmosensory areas [64–66]. The presence of aquaporin water channels would substantially increase the hydraulic conductivity of the blood-brain barriers where they are expressed and probably play an important role in the formation of oedema.

7 Other solute transporters of the blood-brain barriers

7.1 Glucose

The transport of both glucose and amino acids at the BBB has recently been reviewed [67]. There are at least 11 members of the family of facilitated glucose transporters GLUT [67, 68]. Glucose transport across the BBB is mainly facilitated by the GLUT1 transporter which is insulin insensitive. GLUT 1 is expressed in both the luminal and abluminal membranes of the cerebral endothelial cells [69–72] and facilitates glucose entry into the CNS down a concentration gradient from plasma to brain.

The continuous utilisation of glucose by brain tissue produces a sink phenomenon maintaining the concentration gradient between blood and brain. However, the presence of the GLUT1 transporter in both luminal and abluminal membranes is not sufficient in itself to fully explain a maintained directional transport of glucose from blood to brain under all conditions, where possible altered diffusion gradients may occur resulting from rapid fluctuations in glucose levels on both sides of the BBB. Because of high glucose utilization by the brain and the absence of paracellular pathways, cerebral interstitial fluid glucose levels are on a "knife-edge" and become very susceptible to falls in plasma glucose which could easily cause the interstitial fluid glucose concentration to plummet. It has been proposed that a localisation and high activity of hexokinase in a cytoplasmic compartment immediately under the luminal membrane of the cerebral endothelial cells is probably critical for an adequate brain glucose supply maintaining a net cerebral influx of glucose with a minimal concentration gradient [73]. The K_m of GLUT1 is approximately 6–8 mM [73], close to blood glucose levels under normoglycaemic conditions. The concentration of glucose in brain interstitial fluid is estimated to be between 0.5–3.5 mM [74] and is similar in the CSF.

The glucose transporter of neurons is predominately GLUT3, with GLUT1 (45 kDa isoform) expression on glia and GLUT5 on microglia [75]. Also significantly the insulin-sensitive GLUT4 may be expressed in the hypothalamus [76]. Interestingly the sodium-dependant glucose transporter SGLT1 appears to be expressed at the abluminal membranes of the BBB [77] also perhaps significantly expressed alongside the sodium/potassium ATPase [57], where it is well positioned to transport a quantity of glucose back into the endothelial cells of the BBB. The physiological significance of this abluminal SGLT1 expression is not clear [77]. The choroid plexus expresses GLUT1 (45 kDa isoform) exclusively in the epithelial cells [78].

7.2 Amino acids

The blood-brain barrier also contains a spectrum of amino acid carriers, which allow the brain access to amino acids which are polar molecules and would otherwise be excluded by the barrier [79–81].

The amino acid transport mechanisms are also important regulatory mechanisms controlling the amino acid composition of the brain extracellular fluid and supplying the brain with its amino acid requirements for protein synthesis and repair. As with other body tissues several amino acids are essential to the brain by virtue of the fact that they cannot be fully synthesised *de novo*, although the precise essential dietary requirements show some species variation. Also, the synthesis of some neurotransmitters such as serotonin, histamine and dopamine which are derived from amino acids may potentially become limited by the supply of precursor [82, 83]. The amino acid transporters are asymmetrically expressed in the luminal and abluminal membranes of the cerebral endothelial cells which imparts directionality to their transport functions. This polarity of expression, along with that of many other transporters is induced in part by soluble factors and part by an intimate proximity with glial end feet.

The BBB disposition of amino acid transporters and their typical substrates are shown in Figure 8. Two of the transporters, L and y^+, are Na^+ independent and are facilitative and bi-directional in nature and the others are driven by the sodium gradient and operate as co-transporters. The sodium independent transporters are the system for large neutral amino acids (LNAA), the L system, which is largely expressed as the high affinity L_1 isoform in the CNS and the system for cationic amino acids system y^+. These transporters are expressed at a high level in both the luminal and abluminal cell membranes and can transport their substrates across the BBB from blood to brain. The Na^+ dependent amino acid transporters, systems A, $B^{o,+}$, ASC, X^-_{AG} and system β for β amino acids, have a high affinity for their substrates. Systems A, ASC and $B^{o,+}$ are expressed predominantly in the abluminal membrane of the cerebral endothelial cells whereas systems X^-_{AG} and β are expressed in both luminal and abluminal membranes [67, 81].

System L will transport large neutral amino acids with hydrophobic branched or aromatic side chains, the more hydrophobic the side-chain the greater the affinity of an amino acid for the transporter (defined as $1/K_m$) [84]. The L system transporter is composed of a heterodimer of two proteins either LAT1 or LAT2 and 4F2hc a ubiquitous cell surface antigen (CD98). These subunits are covalently linked via a disuphide bond [85, 86]. 4F2hc in this situation is suggested to be acting both a cell trafficing protein, directing the transporter into the cell membrane and possibly also a scaffolding protein orientating and supporting the LAT transporter in the membrane.

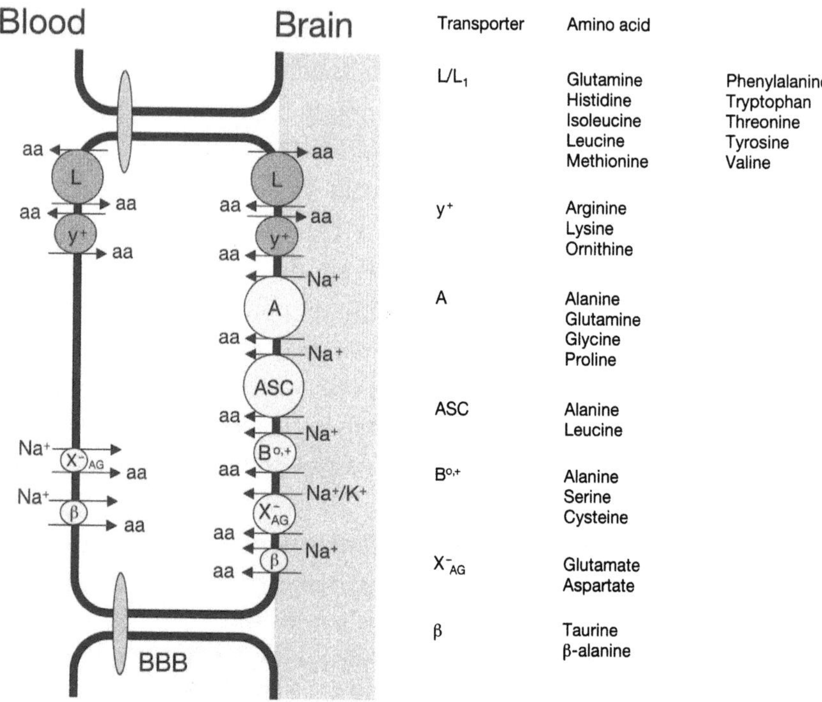

Figure 8.
Amino acid transport at the BBB. The expression of amino acid transporter in the BBB shows polarity with different transporters expressed in the luminal and abluminal membranes. The level of expression of some amino acid transporters also differs between the two membranes. The diameter of the circles indicating the transporter has been drawn approximately equal to the believed level of expression and activity of the relevant transporter.

L(L1), large neutral amino acid transporter; y+, cationic amino acid transporter; A(ATA), amino acid transporter (preferring small neutral amino acids and an N-methyl group); ASC, neutral amino acid transporter (preferring alanine, serine and cysteine); Bo,+, neutral/cationic amino acid transporter; X-AG, amino acid transporter (preferring glutamic acid and aspartic acid); β, β-amino acid transporter.

The transport proteins forming system y+, and also system ASC, appear also to require this association with a 4F2hc subunit to be functional in the cell membrane [67, 86]. System y+, also termed CAT1, transports the essential cationic amino acids arginine, lysine and also ornithine into brain. y+ (CAT1) mRNA is enriched 38% in brain capillaries compared to whole brain [87].

System A is expressed in all tissues and there are three reported isoforms ATA1-3, with ATA1 and ATA2 the principal isoforms expressed in the BBB [88, 89].

Both system A and ASC are located in the abluminal membrane of the endothelial cells and have a preference for small neutral amino acids [90-95].

System $B^{o,+}$ has affinity for both neutral and basic amino acids [96] and system β transports β-alanine and taurine in both the luminal and abluminal cell membranes [95, 97].

The enzyme γ-glutamyl transpeptidase is expressed in the luminal membrane of the brain capillary endothelial cells and is a key enzyme in the γ-glutamyl cycle. It removes the γ-glutamyl moiety from glutathione (GSH) which is then transported into the endothelial cell where it is converted into oxoproline. Oxoproline is then thought to stimulate the activity of system A principally, and also system $B^{o,+}$ in the abluminal membrane of the cell. The oxoproline is further converted into glutamate and re-incorporated into GSH in the cell [91]. The γ-glutamyl cycle thus plays a part in regulating amino acid transport in the endothelial cells and provides an intracellular generator of GSH which can be exchanged for organic anions by oatp transporters (see below).

The anionic amino acids are transportered by system X^-_{AG} which is expressed in both the luminal and abluminal endothelial cell membranes with a greater activity on the abluminal membrane. System X^-_{AG} will transport both glutamate and aspartate [98–101]. As the result of the assymetry of expression of X^-_{AG} net efflux of glutamate from the brain would appear to be 20-fold higher than influx [82]. The presence of system X^-_{AG} transporting glutamate and asparatate and also system ASC transporting glycine into the BBB endothelium from adjacent brain interstitial fluid is an important component, in combination with re-uptake by neurones and glia and transport from CSF to blood across the choroid plexus, in maintaining a low level of these excitatory neurotransmitters in brain extracellular fluid.

In the choroid plexus a number of amino acid transporters are expressed [102]. Studies in the perfused sheep choroid plexus with tracer levels of amino acid have suggested that there is a net entry of amino acid, from blood to CSF across the plexus epithelium [103, 104]. A carrier-mediated uptake of a number of amino acids from the CSF in rabbits has been shown [105] and this has been generally assumed to be due to transport from CSF to blood, largely across the choroid plexus. However, it would appear that under physiological conditions, there is a net entry of amino acids to CSF across the choroid plexus and that the overall low level of amino acids in CSF compared to blood is the result of uptake and utilization by brain tissue.

7.3   Nucleosides, nucleotides, and nucleobases

The brain ultimately acquires most of the nucleotides that it requires for DNA and RNA synthesis from blood as the adult mammalian brain does not have a significant capacity for the *de novo* synthesis of these precursors [106]. Salvage pathways of the nucleobases hypoxanthine and adenine are also and important source of nucleotides [106, 107].

Early brain uptake index (BUI) studies [108] demonstrated rapid and saturable nucleoside transport into brain after intracarotid injection. The nucleoside transporters are grouped into two classes the equilibrative nucleoside transporters (ENT) and the concentrative nucleoside transporters (CNT). There are two equilibrative transporters distinguished by their sensitivity to the inhibitor nitrobenzylmercaptopurine riboside (NBMPR). The equilibrative transporter ENT1(es) is sensitive to this inhibitor whilst ENT2 (ei) is not and both transporters will accept both purine and pyrimidine nucleosides [109]. There are at least 6 subtypes of concentrative transporters (CNT1-6), all of which are Na^+ dependent and function as co-transporters using the sodium gradient across the cell membrane to move their substrates against a concentration gradient [109]. Of the concentrative transporters CNT1(cif) and CNT2(cit) are present in the BBB and CNT3(cib) is expressed in the choroid plexus together with both equilibrative transporters [109]. Recent *in vitro* experiments using immortalized rat brain endothelial cells which are free from contamination with other cell types support this distribution of transporter with the demonstration of the functional expression of ENT1, ENT2, CNT1 and CNT2 [110]. CNT1 prefers purines, and CNT2 pyrimidines whereas CNT3 will accept both classes of nucleosides [109]. The nucleobase transporter SVCT2 is also present in brain [68].

7.4   Monocarboxylic acid transporters

The BBB contains transporters for monocarboxylic acids. Seven subtypes of monocarboxylic acid transporters are so far recognized (MCT1-7) [68]. The monocarboxylic acids transporters are proton coupled, bi-directional, and are capable of transporting lactate, pyruvate and a number of other short-chain monocarboxylic acids, e.g. acetate and the ketone bodies acetoacetate and γ-hydroxybutryrate, into brain which can be utilized as metabolic substrates [111]. In the brain MCT1 and MCT2 appear to be expressed at significant levels [112] and both are capable of transporting lactate. MCT1 is expressed in the endothelial cells of the BBB in both the luminal and abluminal membranes [111–113] whereas MCT2 is expressed in the abluminal membranes of the endothelial cells and on the foot processes of astrocytes [7, 105].

It is suggested that MCT1 may have a higher affinity for moncocarboxylic acids which can serve as brain metabolites such as ketone bodies [111] and MCT2 may be significant in the removal of lacate from brain to blood by maintaining a high concentration in the interstitial fluid close to the abluminal membrane of the endothelial cells where it can then be removed to blood by MCT2 [7, 113].

7.5   Organic anion transporters

At least four members of this family of organic anion transporters have been described. OAT1 and 2 are principally expressed in liver and OAT3 in the brain [114]. The members of the family are multispecifc for their substrates, each transporting a wide range of structures but each of thetransporters has specific substrate preferences. All members of the family appear to transport para-aminohippuric acid (PAH) [114]. As a group they transport various low molecular weight and weakly amphiphilic substrates [115] and their physiological substrates are probably dicarboxylates.

Based on functional studies [116, 117] OAT3 would appear to be the major OAT present in the BBB and the choroid plexus epithelium and also shows a strong signal with reverse transcription-polymerase chain reaction (RT-PCR) and Northern blotting from these locations [114]. OAT1, 2 and 3 are also present in the apical membranes of the choroid plexus epithelium [118, 119]. The OAT transporters are exchange transporters and will exchange an organic anion for either α-ketoglutarate (OAT1) [120] or Cl$^-$ or HCO$_3^-$ in the case of OAT3 [121, 122].

7.6   Organic anion transporting polypeptides

The organic anion transporting polypeptides (oatps in rat/OATPs in human) are a large family of transporters handling a wide range of substrates which are anionic at physiological pH [123]. Their physiological substrates include bile salts, steroid conjugates, thyroid hormones, leucotriene C4, biotin, glutathione, opiate and other small peptides and peptidomimetics [124–130].

Their transport is driven by the substrate concentration gradient and they function as either organic anion/HCO_3^- exchangers or organic anion/gluathione exchangers, the transport is driven by the substrate concentration and is thus reversible depending on the direction of the concentration gradient [126, 131].

Studies in the rat suggest that oatp1 is expressed in the apical membrane of the choroid plexus epithelial cells [132] and oatp2 in the basolateral membranes of these cells and in the BBB endothelial cells on both the luminal and abluminal membranes [129, 133]. Human OATP-B appears to be the principle human brain transporter of this group and is expressed at the BBB [133, 134].

7.7 Organic cation transporters (OCT and OCTN)

Three organic cation transporters OCT 1-3 are described in the rat and the human and the transporters appear structurally to be closely related and transport a variety of organic cations [123]. They are driven by the cell membrane potential and probably function as sodium independent proton/organic ion exchangers [135]. Relatively little is known about their distribution in the brain. In the rat OCT3 is expressed in brain tissue generally [136] and OCT2 and OCT3 are expressed in the apical membrane of the choroid plexus [135]. The interesting observation has been made [137] that quinacrine is taken up at the apical membrane of the choroid plexus by OCT2 but appears to be released at the basal membrane by a process of exocytosis.

Two further organic cation transporters have been characterised as OCTN1 and OCTN2, the N in the acronym referring to "novel". OCTN transporters have been described in the human, rat and mouse and appear fairly widespread between species [138]. They are structurally distinct from the OCTs but within the group are closely structurally related [123] and also function as sodium independent, bi-directional, organic anion proton exchangers [139–141].

OCTN1, 2 and OCTN3 are expressed in cerebral endothelial cells in the mouse [142] whereas OCTN2 is expressed in the endothelium of the bovine, porcine and human BBB [143]. No human equivalent of mouse OCTN3 has yet been described [143]. OCTN2 appears to be expressed in the luminal membrane of the cerebral endothelial cells and both OCTN1 and 2, or an additional unknown transporter at the abluminal membrane [142]. Both

OCTN1 and OCTN2 are expressed in the choroid plexus but the location of expression remain to be determined [119].

8   ABC transporters of the blood-brain barriers

The ABC transporters are a large group of structurally related transporters some of which play a critcal role in the blood-brain barrier by functioning as active efflux pumps transporting a wide range of structurally diverse compounds out of the CNS. The ABC transporters are able to hydrolyse ATP and possess two ATP binding sites on each transporter molecule. The hydrolysis of ATP provides the energy for their activity in moving their substrates against a concentration gradient and/or opposing their natural tendancy to partition into the lipid of the cell membrane. The ability of the ABC transporters to bind and hydrolyse ATP gives rise to the acronym ABC, indicating the ATP-binding cassette of transporters. At present in humans about 30 ABC transporters are recognised and are grouped into four families [144]. Of recognised importance in the blood-brain barrier are P-glycoprotein (Pgp), multidrug resistance-associated protein (MRP) and breast cancer-resistance protein (BCRP) each of which are members of a separate ABC family [144]. Their functions in the blood-brain and blood-CSF barriers have recently been the topic of some general reviews [121–124, 145, 146] and they will only be briefly reviewed here. The substrates of these ABC transporters include a number of physiological substrates, potentially neurotoxic xenobiotics and a large number of therapeutic drugs. This latter phenomenon presenting a huge challenge to effective brain drug delivery and to the treatment of CNS disease. The function of the ABC transporters in the CNS are generally as detoxifying and neuroprotective systems for the brain.

8.1   P-glycoprotein (Pgp)

P-gycoprotein is encoded by a group of related genes in mammals called multidrug resistance genes [147]. In humans there are two gene products MDR1 and MDR2 [148] and in rodents three mdr1a, mdr1b and mdr2 [149]. Of significance in the CNS are MDR1, mdr1a and mdr1b, for simplicity collectively termed here Pgp. MDR1 and mdr1a are expressed in the luminal membrane of the cerebral endothelial cells where they can transport their substrates directly into blood [150–154]. The substrates of Pgp are generally large planar molecules which are lipophilic and many steroids and glucuronides are endogenous substrates for Pgp [123]. Pgp is also expressed at the apical surface of the choroid plexus [155]. The location is described as sub-apical and appears to be largely in vesicles immediately under and in close association with the apical epithelial membrane. This may represent an intracellular pool of Pgp that can be inserted into the cell membrane or may suggest that the Pgp substrates are sequestered into vesicles in a manner reminiscent of OCT2, see above [137]. In rodents, mdr1b does not appear to be principally expressed in capillary endothelial cells but appears to be expressed generally in brain parenchyma [156].

8.2 Multidrug resistance-associated proteins (MRP)

Seven isoforms of MRP are currently recognised [157], MRP1 to MRP7. In the rat mrp2 is expressed in the luminal membranes of the BBB [158, 159] and RT-PCR signals for mrp1, mrp4 and mrp5 and a very weak signal for mrp6 are present in freshly isolated bovine brain capillaries [160]. Functional studies have suggested that MRP1 is not present on the luminal surface of the brain endothelium [161]. This finding is also consistent with the observation that in the triple knockout mouse, where mdr1a, mdr1b and mrp1 are absent, that there is no increase in etoposide brain distribution although it is a substrate for all three transporters [162]. Putting all of these observations together they would suggest that mrp1 is expressed in the abluminal and basolateral membranes of the blood-brain barrier and the choroid plexus respectively and mrp2 in the luminal membrane of the blood-brain barrier. The location of expression in the CNS and the substrate preferences for mrp4, mrp5 and mrp6 are still open to question.

Substrate preferences for MRPs appear to overlap to some extent with Pgp and include glucuronides, cAMP, cGMP, glutathione conjugates, the bile acids glycocholate and taurocholate and organic anions are substrates [123]. MRP shows a preference for compounds containing some charge at physiological pH as demonstrated by the organic anions transported.

8.3 Breast cancer-resistance protein (BCRP)

Recently breast cancer-resistance protein (BCRP) sometimes called mitoxantrone-resistance protein (MXR1) has been shown to be present in the blood-brain barrier [163–165]. The protein is clearly shown to be expressed in the luminal membrane of the cerebral endothelial cells [164] and presumably functions in a similar manner to Pgp. A number of drugs have been shown to be substrates for BCRP [123] but its physiological substrates as yet are unknown. It is of interest that comparing the structures of Pgp, MRP and BCRP, that BCRP appears to be a half molecule with only one ATP binding site compared to the other two transporters and is thus assumed to dimerise in order to be active [166].

9   Specific peptide transporters

It is now well established that many peptides are transferred across the BBB by means of specific transport mechanisms [167–171]. The PEPT2 transporter has been shown to be present in the choroid plexus [172] and glia [173] but is absent at the BBB. The routes and the mechanisms by which systemically circulating peptides may enter the brain and influence CNS function will not be reviewed here as the topic forms a subsequent chapter in this volume.

10   Metabolic conversion in the blood-brain barriers

It is now apparent that both the BBB and the BCSFB in addition to physical and transport barriers are also metabolic barriers to transport [174–176]. Many drugs and solutes are metabolized as they pass through the BBB and BCSFB. The barriers contain active Phase 1 enzymes (cytochromes P450, flavin-dependent oxygenases, monoamine oxidases, reductases, hydrolases) and Phase 2 enzymes (conjugating enzymes, sulphotransferases, GSH-transferases, UDP-glucuronosyltransferases) [123, 174–176]. Metabolism and/or conjugation may occur prior to subsequent efflux transport by an ABC transporter.

There are also a number of significant peptidases in the blood-brain barriers and in brain tissue. Angiotensin converting enzyme (ACE) is present at the luminal surface of the BBB and in the microvilli at the apical surface of the choroidal epithelium [177]. Aminopeptidases and endopeptidases are also present as membrane bound enzymes on the cell membranes of neurons and glia and some exist in soluble form in brain extracellular fluid [178–182]. These enzymes may function to limit peptide entry and their subsequent action within the CNS.

11 Conclusion

The advent of molecular biological techniques have increased our understanding of the blood brain barriers to an enormous extent in recent years. These advances, combined with the more traditional structural, *in vitro* and *in vivo* techniques, together provide very powerful tools with which to understand barrier function. Far from being regarded as simple physical barriers to solute diffusion, the blood-brain barriers are now being fully appreciated as physiological, metabolic and transport interfaces, which react and interact with a host of blood-borne factors and cellular elements.

References

1   Kniesel U, Wolburg H (2000) Tight junctions of the blood-brain barrier. *Cell Mol Neurobiol* 20: 57–76

2   Wolburg H, Wolburg-Buckholz K, Liebner S, Engelhardt B (2001) Claudin-1, claudin-2 and claudin 11 are present in tight junctions of choroid plexus epithelium of the mouse. *Neurosci Lett* 307: 77–80

3   Begley DJ (1996) The blood-brain barrier: principles for targeting peptides and drugs to the central nervous system. *J Pharm Pharmacol* 48: 136–146

4   Levin VA (1980) Relationship of octanol/water partition coefficient and molecular weight to rat brain capillary permeability. *J Med Chem* 23: 682–684

5   Sugiyama Y, Kusuhara H, Suzuki H (1999) Kinetic and biochemical analysis of carrier-mediated efflux of drugs through the blood-brain and blood-cerebrospinal fluid barriers: importance in the drug delivery to the brain. *J Control Rel* 62: 179–186

6   Tamai I, Tsuji A (2000) Transport-mediated permeation of drugs across the blood-brain barrier. *J Pharm Sci* 89: 1371–1388

7   Mertsch K, Maas J (2002) Blood-brain barrier penetration and drug development from an industrial point of view. *Curr Med Chem (Central Nervous System Agents)* 2: 187–201

8   Reese TS and Karnovsky MJ (1967) Fine structural localization of a blood-brain barrier to exogenous peroxidase. *J Cell Biol* 34: 207–217

9   Brightman MW (1968) The intracerebral movement of proteins injected into blood and cerebrospinal fluid of mice. *Prog Brain Res* 29: 19–37

10   Brightman MW and Reese TS (1969) Junctions between intimately apposed cell membranes in the vertebrate brain. *J Cell Biol* 40: 648–677

11   Butt AM (1995) Effect of inflammatory agents on electrical resistance across the blood-brain barrier in pial microvessels of anaesthetized rats. *Brain Res* 696: 145–150

12   Brightman MW, Tao-Cheng J-H (1993) Tight junctions of brain endothelium and epithelium. In: WM Pardridge (ed): *The blood-brain barrier cellular and molecular biology*. Raven Press, 107–125

13   Morita K, Sasaki H, Furuse M, Tsukita S (1999) Endothelial claudin: claudin 5 TMVCF, constitutes tight junction strands in endothelial cells. *J Cell Biol* 147: 185–194

14   Furuse M, Hirase T, Itoh M, Nagafuchi A, Yonemura S, Tsukita S, Tsukita S (1993) Occludin: a novel integral membrane protein localizing at tight junctions. *J Cell Biol* 123: 1777–1788

15   Saitou M, Fujimoto K, Doi Y, Itoh M, Fujimoto T, Furuse M, Takano H, Noda T, Tsukita S (1998) Occludin-deficient embryonic stem cells can differentiate into polarized epithelial cells bearing tight junctions. *J Cell Biol* 141: 397–408

16   Del Maschio A, de Luigi A, Marin-Padura I, Brockhaus M, Bartfai T, Fruscella P, Adorini L, Martino G, Furlan R, de Simoni MG et al (1999) Leukocyte recruitment in the cerebrospinal fluid of mice with experimental meningitis is inhibited by an antibody to junctional adhesion molecule (JAM). *J Exp Med* 190: 1351–1356

17   Cordonensi M, D'Atri F, Hammar E, Parry DAD, Kendrick-Jones J, Shore D, Citi S (1999) Cingulin contains lobular and coiled-coil domains and interacts with ZO-1, ZO-2, ZO-3 and myosin. *J Cell Biol* 147: 1569–1581

18   Martinez-Palomo A, Erlij D (1975) Structure of tight junctions in epithelia with different permeability. *Proc Natl Acad Sci USA* 72: 4487–4491

19   Stevenson BR, Anderson JM, Goodenough DA, Mooseker MS (1988) Tight junction structure and ZO-1 content are identical in two strains of Madin-Darby canine kidney cells which differ in transepithelial resistance. *J Cell Biol* 107: 2401–2408

20   Yuan Y, Meng FY, Huang Q, Hawker J, Mac Wu H (1998) Tyrosine phosphorylation of paxillin/pp125FAK and microvascular endothelial barrier function. *Am J Physiol* 275: (*Heart Circ Physiol* 44): H84–H93

21   Staddon JM, Herrenknecht K, Smales C, Rubin LL (1995) Evidence that tyrosine phosphorylation may increase tight junction permeability. *J Cell Sci* 108: 609–619

22   Fanning AS, Jameson BJ, Jesaitis LA, Anderson JM (1998) The tight junction protein ZO-1 establishes a link between the transmembrane protein occludin and the actin cytoskeleton. *J Biol Chem* 273: 29745–29753

23   Huber JD, Egleton RD, Davis TP (2001) Molecular physiology and pathophysiology of tight junctions in the blood-brain barrier. *Trends in Neurosci* 24: 719–725

24   Trentin AG, Alvarez-Silva M, Moura Neto V (2001) Thyroid hormone induces cerebellar astrocytes and C6 glioma cells to secrete mitogenic growth factors. *Am J Physiol* 281: E1088–1094

25   Deli MA, Descamps L, Dehouck M-P, Ceccheli R, Joo F, Abraham CS, Torpier G (1995) Exposure of tumor necrosis factor to luminal membrane of bovine brain capillary endothelial cells cocultured with astrocytes induces a delayed increase of permeability and cytoplasmic stress fiber formation of actin. *J Neurosci Res* 41: 717–726

26   Ozaki H, Ishii K, Horiuchi H, Arai H, Kawamoto K, Okawa A, Iwamatsu A, Kita T (1999) Cutting edge: combined treatment of TNF-alpha and IFN-gamma causes redistribution of junctional adhesion molecule in human endothelial cells. *J Immunol* 163: 553–557

27   Hickey W (1999) Leukocyte traffic in the central nervous system: the participants and their roles. *Semin Immunol* 11: 125–137

28   Wong D, Prameya, R, Dorovini-Zis K (1999) *In vitro* adhesion and migration of T lymphocytes across monolayers of human brain microvesel and endothelial cells: regulation by ICAM-1, VCAM-1, E-selectin and PECAM-1. *J Neuropathol Exp Neurol* 58: 138–152

29   Burns AR, Bowden RA, MacDonell SD, Walker DC, Odebunmi TO, Donnachie EM, Simon SI, Entman ML, Smith CW (2000) Analysis of tight junctions during neutrophil transendothelial migration. *J Cell Sci* 113: 45–57

30   Faustman PM, Dermietzel R (1985) Extravasation of polymorphonuclear leukocytes

from the cerebral microvasculature. Inflammatory response induced by alpha-bungaro-toxin. *Cell Tissue Res* 242: 399–407

31   Sixt M, Engelhardt B, Pausch F, Hallman R, Wendler O, Sorokin CM (2001) Endothelial cell laminin isoforms, laminins 8 and 10, play decisive roles in T cell recruitment across the blood-brain barrier in experimental autoimmune encephalomyelitis. *J Cell Biol* 153: 933–946

32   Raub TJ, Newton CR. (1991) Membrane recycling, adsorptive and receptor-mediated endocytosis by bovine cerebral microvessel endothelial cell monolayers. In: G Wilson, A Zweibaum, L Illum, S Davis (eds): *Pharmaceutical Applications of Cell and Tissue Culture to Drug Transport.* Plenum Press, New York, 203–216

33   Villegas JC, Broadwell RD (1993) : Transcytosis of protein through the mammalian cere-bral epithelium and endothelium. II Adsorptive transcytosis of WGA–HRP and the blood-brain and brain-blood barriers. *J Neurocytol* 22 : 67–80.

34   Vorbrodt AW (1989) Ultracytochemical characterization of anionic sites in the wall of brain capillaries. *J Neurocytol* 18 : 359–368

35   Pardridge WM, Triguero D, Buciak J, Yang J (1990) Evaluation of cationized rat albumin as a potential blood-brain barrier drug transport vector. *J Pharmacol Exp Ther* 255: 893–899

36   Pardridge WM, Eisenberg J, Cefalu WT (1985) Absence of albumin receptor on brain cap-illaries *in vivo* or *in vitro*. *Am J Physiol* 249 (3 Pt 1): E264–E267

37   Schnitzer JE (1992) gp60 is an albumin-binding glycoprotein expressed by continuous endothelium involved in albumin transcytosis. *Am J Physiol* 262 (*Heart Circ Physiol* 31): H246–H254

38   Friden PM, Walus LR, Musso GF, Taylor MA, Malfroy B, Starzyk RM (1991) Anti-trans-ferrin receptor antibody and antibody-drug conjugates cross the blood-brain barrier. *Proc Natl Acad Sci USA* 88: 4771–4775

39   Morgan EH, Moos T (2002) Mechanism and developmental changes in iron transport across the blood-brain barrier. *Dev Neurosci* 24: 106–113

40   Zhang Y, Pardridge WM (2001) Neuroprotection in transient focal brain ischemia after delayed intravenous administration of brain-derived neurotropic factor conjugated to a blood-brain barrier drug targeting system. *Stroke* 32: 1378–1384

41   Broadwell RD, Brightman MW (1976) Entry of peroxidase into neurons of the central and peripheral nervous system from extracerebral and cerebral blood. *J Comp Neurol* 166: 257–284

42   Fiala M, Looney DJ, Stins M, Way DD, Zhang X, Gan F, Chiappelli F, Schweitzer ES, Shap-shak P, Weinand M et al (1997) TNF-alpha opens a paracellular route for HIV-1 invasion across the blood-brain barrier. *Mol Med* 3: 553–564

43   Banks WA, Freed EO, Wolf KM, Robinson SM, Franko M, Kumar VB (2001) Transport of human immunodeficiency virus type 1 pseudoviruses across the blood-brain barrier: role of envelope proteins and adsorptive endocytosis. *J Virol* 75: 4681–4691

44   Brightman MW, Ishihara S, Chang L (1995) Penetration of solutes viruses and cells across the blood-brain barrier. *Curr Topics Microbiol Immun* 202: 63–78

45   Broadwell RD, Balin BJ (1985) Endocytic and exocytic pathways of the neuronal secre-tory process and trans-synaptic transfer of wheat germ agglutinin-horseradish peroxi-dase *in vivo*. *J Comp Neurol* 242: 632–650

46   Rinaman L, Roesch MR, Card JP (1999) Retrograde transynaptic pseudorabies virus infection of central autonomic circuits in neonatal rats. *Develop Brain Res* 114: 207–216

47   Frey WH, Liu J, Chen X, Thorne RG, Fawcett JR, Ala TA, Rahman Y-E (1997) Delivery of
      125I-NGF to the brain *via* the olfactory route. *Drug Delivery* 4: 87–92

48   Sakane T, Akizuki M, Taki Y, Yamashita S, Sezaki H, Nadai T (1995) Direct drug transport
      from the rat nasal cavity to the cerebrosoinal fluid: the relation to the molecular weight
      of drugs. *J Pharm Pharmacol* 47: 379–381

49   Pardridge WM (1991) *Peptide Drug Delivery to the Brain*. Raven Press. New York, 357

50   Bradbury M, Cserr HF, Westrop RJ (1981) Drainage of cerebral interstitial fluid into deep
      cervical lymph of the rabbit. *Am J Physiol* 240: F329–F336

51   Jarolim KL, McCosh JK, Howard MJ, John DT (2000) A light microscopy study of the
      migration of *Naegleria fowleri* from the nasal submucosa to the central nervous system
      during the early stage of primary amebic meningoencephalitis in mice. *J Parasitol* 86:
      50–55

52   Knopf PM, Cserr HF, Nolan SC, Wu T-Y, Harling-Berg CJ (1995) Physiology and immunol-
      ogy of lymphatic drainage of interstitial and cerebrospinal fluid from the brain. *Neu-
      ropathol Appl Neurobiol* 21: 175–180

53   Nadal A, Fuentes E, Pastor J, McNaughton PA (1995) Plasma albumin is a potent trigger
      of calcium signals and DNA synthesis in astrocytes. *Proc Natl Acad Sci USA* 92: 1426–1430

54   Tao-Cheng J-H, Nagy Z, Brightman MW (1987) Tight junctions of brain endothelium *in
      vitro* are enhanced by astroglia. *J Neurosci* 7: 3293–3299

55   Westergaard E, Brightman MW (1973) Transport of proteins across normal cerebral arte-
      rioles. *J Comp Neurol* 152: 17–44

56   Brightman MW (1999) Blood-brain barrier: penetration by solutes and cells. In: G Adel-
      man and BH Smith (eds): *Encyclopedia of Neuroscience*, 2nd ed., Elsevier Science, Amster-
      dam, 241–245

57   Betz AL, Firth JA, Goldstein GW (1980) Polarity of the blood-brain barrier: Distribution
      of enzymes between the luminal and abluminal membranes of brain capillary endothe-
      lial cells. *Brain Res* 192: 17–28

58   Zeuthan T, Wright EM (1981) Epithelial potassium transport: tracer and electrophysio-
      logical studies in choroids plexus. *J Membrane Biol* 60: 105–128

59   Johansen C E, Swenney SM, Parmalee JT, Epstein MH (1990) Cotransport of sodium and
      chloride by the adult mammalian choroid plexus. *Am J Physiol* 258: C211–C215

60   Speake T, Whitwell C, Kajita H, Majid A, Brown PD (2001) Mechanisms of CSF secretion
      by the choroid plexus. *Microsc Res Tech* 52: 49–52

61   Nielsen S, Smith BL, Christensen EI, Agre P (1993) Distribution of the aquaporin CHIP
      in secretory and resorptive epithelia and capillary endothelia. *Proc Natl Acad Sci USA* 90:
      7275–7279

62   Wu Q, Delpire E, Hebert SC, Strange K (1998) Functional demonstration of Na^+-K^+-$2Cl^-$
      cotransporter activity in isolated, polarizedchoroid plexus cells. *Am J Physiol* 275:
      C1565–1572

63   Sobue K, Yamamoto N, Yoneda K, Fujita K, Miura Y, Asia K, Tsuda T, Katsuya H, Kato T
      (1999) Molecular cloning of two bovine aquaporin-4 cDNA isoforms and their expres-
      sion in brain endothelial cells. *Biochim Biophys Acta* 1489: 393–398

64   Nielsen S, Nagelhus EA, Amiry-Moghadden M, Bourque C, Ottesen OP (1997) Spe-
      cialised membrane domains for water transport in glial cells: High reolution immuno-
      gold cytochemistry of aquaporin-4 in rat brain. *J Neurosci* 17: 171–180

65   Nico B, Frigeri A, Nicchia GP, Quondamatteo F, Herken R, Errede M, Ribatti D, Svelto M,

Roncali L (2001) Role of aquaporin-4 water channel in the integrity of the blood brain barrier. *J Cell Sci* 114: 1297–1307

66   Badaut, J, Lasbennes F, Magistretti PJ, Luca R (2002) Aquaporins in Brain : Distribution, Physiology and Pathophysiology. *J Cereb Blood Flow Metab* 22: 367–378

67   Mann GE, Yudilevich DL, Sobrevia L (2003) Regulation of amino acid and glucose transporters in endothelial and smooth muscle cells. *Physiol Rev* 83: 183–252

68   Zhang EY, Knipp GT, Ekins S, Swann PW (2001) Structural biology and function of solute transporters: implications for identifying and designing substrates. *Drug Metab Rev* 34: 709–750

69   Farrel CL, Pardridge WM (1991) Blood-brain glucose transporter is asymmetrically distributed on brain capillary endothelial lumenal and ablumenal membranes: an electron microscope study. *Proc Natl Acad Sci USA* 88: 5779–5783

70   Pardridge W M, Boado RJ (1993) Molecular cloning and regulation of gene expression of blood brain barrier glucose transporter. In: WM Pardridge (ed): *The blood-brain barrier*. Raven Press, New York, 395–440

71   Cornford EM, Hyman S, and Swartz BE (1994) The human brain GLUT1 glucose transporter: Ultrastructural localization to the blood-brain barrier endothelia. *J Cereb Blood Flow Metab* 14: 106–112

72   Cornford EM, Hyman S, Cornford ME, Landaw EM, Delgado-Escueta AV (1998) Interictal seizure resections show two configurations of endothelial GLUT1 glucose transporter in the human blood brain barrier. *J Cereb Blood Flow Metab* 18: 26–42

73   McAllister MS, Krizanac-Bengez L, Macchia F, Naftalin RJ, Pedley KC, Mayberg MR, Marroni M, Leaman S, Stanness KA, Janigro D (2001) Mechanisms of glucose transport at the blood-brain barrier: an *in vitro* study. *Brain Res* 904: 20–30

74   Drewes LR (1998) Biology of the blood-brain glucose transporter. In: WM Pardridge (ed): *Introduction to the blood-brain barrier*. Cambridge University Press, Cambridge, 165–174

75   Maher F, Vannuchi S, Simpson IA (1993) Glucose transporter proteins in brain. *FASEB J* 8: 1003–1011

76   Livingstone C, Lyall H, Gould GW (1995) Hypothalamic Glut 4 expression: A glucose and insulin sensing mechanism. *Mol Cell Endocrinol* 107: 67–70

77   Lee WJ, Peterson DR, Sukowski EJ, Hawkins RA (1997) Glucose transport by isolated plasma membranes of the bovine blood-brain barrier. *Am J Physiol Cell Physiol* 272: C1552–C1557

78   Vannuchi SJ (1994) Developmental expression of GLUT1 and GLUT4 glucose transporters in brain. *J Neurochem* 62: 240–246

79   Pardridge WM (1995) Transport of small molecules through the blood-brain barrier; Biology and methodology. *Adv Drug Delivery Rev* 15: 5–36

80   Pardridge WM (1998) Blood-brain barrier carrier-mediated transport and brain metabolism of amino acids. *Neurochem Res* 23: 636–644

81   Smith QR, Stoll J (1998) Blood-brain barrier amino acid transport. In: WM Pardridge (ed): *Introduction to the blood-brain barrier: Methodology, biology and pathology*. Cambridge University Press, Cambridge, 188–197

82   Pardridge WM (1983) Brain metabolism a perspective from the blood-brain barrier. *Physiol Rev* 63: 1481–1535

83   Smith QR, Takasato Y (1986) Kinetics of amino acid transport at the blood-brain barrier studied using an in situ brain perfusion technique. *Ann NY Acad Sci* 481: 186–201

84   Smith QR (1995) Carrier-mediated drug transport at the blood-brain barrier and the

potential for drug targeting to the brain. In: J Greenwood, DJ Begley, MB Segal (eds): *New concepts of a blood-brain barrier*. Plenum Press, New York and London, 265–276

85   Smith QR (2000) Transport of glutamate and other amino acids at the blood-brain barrier. J Nutr 130: 1016S–1022S

86   Kageyama T, Nakamura M, Matsuo A, Yamasaki Y, Takaura Y, Hashida M, Kanai, Y Naito M, TsuroT, Minato N et al (2000) The 4F2hc/LAT1 complex transports L-DOPA across the blood-brain barrier. *Brain Res* 879: 115–121

87   Stoll J, Wadhwani KC, Smith QR (1993) Identification of the cationic amino acid transporter (system y$^+$) of the rat blood-brain barrier. *J Neurochem* 60: 1956–1959

88   Wang H, Huang W, Sugawara M, Devoe LD, Leibach FH, Prasad PD, Ganpathy V (2000) Cloning and functional expression of ATA1 a subtype of amino acid transporter A, from human placenta. *Biochem Biophys Res Comm* 273: 1175–1179

89   Takanaga H, Tokuda S, Ohtsuki S, Hosoya K-I, Terasaki T (2002) ATA2 is predominantly expressed as system A at the blood-brain barrier and acts as brain to blood efflux transport for L-proline. *Mol Pharmacol* 61: 1289–1296

90   Betz AL, Goldstein GW (1978) Polarity of the blood-brain barrier: neutral amino acid transport into isolated brain capillaries. *Science* 202: 225–227

91   Lee WJ, Hawkins RA, Peterson DR, Viña JR (1996) Role of oxoproline in the regulation of neutral amino acid transport across the blood-brain barrier. *J Biol Chem* 271: 19129–19133

92   Lee WJ, Hawkins RA, Viña JR, Peterson DR (1998) Glutamine transport by the blood-brain barrier; a possible mechanism for nitrogen removal. *Am J Physiol* 274: C1101–C1107

93   Sánchez del Pino MM, Hawkins RA, Peterson DR (1995) Neutral amino acid transport charaterization of isolated luminal and abluminal membranes of the blood-brain barrier. *J Biol Chem* 270: 1493–14918

94   Tayarani I, Lefauconnier JM, Roux F, Bourre JM (1987) Evidence for an alanine, serine and cysteine system of transport in isolated brain capillaries. *J Cereb Blood Flow Metab* 7: 585–591

95   Tayarani I, Cloez I, Lefauconnier JM, Bourre JM (1989) Sodium dependent high affinity uptake of taurine by isolated rat brain capillaries. *Biochim Biophys Acta* 985: 168–172

96   Guidotti G G, Gazzola GC (1992) Amino acid stransporters: systematic approach and principles of control. In: MS Kilberg, D Haussinger (eds): *Mammalian amino acid transport*. Plenum Press, New York, 3–30

97   Tamai I, Senmaru M, Teraskai T, Tsuji A (1995) Na$^+$ and Cl$^-$ dependent transport of taurine at the blood-brain barrier. *Biochem Pharmacol* 50: 1783–1739

98   Al-Sarraf H, Preston J, Segal, MB (1995) The entry of acidic amino acids into brain and CSF during development using *in situ* brain perfusion in the rat. *Dev Brain Res* 90: 151–158

99   Al-Sarraf H, Preston J, Segal MB (1997) Changes in the kinetics of the acidic amino acid brain and CSF uptake during development in the rat. *Dev Brain Res* 102: 127–134

100   Benrabh H , Lefauconnier JM (1996) Glutamate is transported across the rat blood-brain barrier by a sodium-independent system. *Neurosci Lett* 210: 9–12

101   Oldendorf WH, Szabo J (1976) Amino acid assignment to one of three blood-brain barrier amino acid carriers. *Am J Physiol* 230: 94–98

102   Segal MB (2001) Transport of nutrients across the choroid plexus. *Micros Res and Tech* 52: 38–48

103   Preston J, Segal MB (1990) The steady-state amino acid fluxes across the perfused choroid plexus of the sheep. *Brain Res* 525: 275–279

104   Preston J, Segal MB (1992) The uptake of anionic and cationic amino acids by the perfused sheep choroid plexus. *Brain Res* 581: 35–355

105   Davson H, Hollingsworth JG, Carey MB, Fenstermacher JD (1982) Ventriculo-cisternal perfusion of twelve amino acids in the rabbit. *J Neurobiol* 13: 293–318

106   Redzic ZB, Segal MB, Gasic JM, Markovic ID, Vojvodic VP, Iskavic A, Thomas SA, Rakic Lj (2001) The characteristics of nucleobase transport and metabolism by the perfused sheep choroid plexus. *Brain Res* 888: 66–74

107   Fox IH, Kelly WH (1978) The role of adenosine and 2'-deoxyadenosine in mammalian cells. *Ann Rev Biochem* 47: 655–686

108   Cornford EM, Oldendorf WH (1975) Inedependant blood-brain barrier transport systems for nucleic acid precursors. *Biochim Biophys Acta* 394: 211–219

109   Lee G, Dallas S, Hong R, Bendayan R (2001) Drug transporters in the central nervous system: brain barriers and brain parenchymal considerations. *Pharmacol Rev* 53: 569–596

110   Chishty M, Begley DJ, Abbott NJ, Reichel A. (2003) Functional characterisation of nucleoside transport in rat brain endothelial cells. *NeuroReport*: 14: 1087–1090

111   Gerhart DZ, Enerson BE, Zhdakina OY, Leino RL, Drewes LR (1997) Expression of monocarboxylate transporter MCT1 by brain endothelium and glia in adult and suckling rats. *Am J Physiol* 273: E207–E213

112   Price NT, Jackson VN, Halestap AP (1998) Cloning and sequencing of four new mammalian monocarboxylic acid transporter (MCT) homlogues confirms the existence of a transporter family with an ancient past. *Biochem J* 329: 321–328

113   Cornford EM, Hyman S (1999) Blood-Brain barrier permeability to small and large molecules. *Adv Drug Deliv Rev* 36: 145–163

114   Kusuhara H, Sekine N, Utsunomiya-Tate M, Tsuda R, Kojima SH, Cha Y, Sugiyama Y, Kanai Y, Endou H (1999) Molecular cloningand charaterisation of a new multispecific organic anion transporter from rat brain *J Biol Chem* 274: 13575–13680

115   Burkhardt G, Wolf NA (2000) Structure of renal organic anion and cation transporters. *Am J Physiol* 278: F853–F866

116   Suzuki H, Sawada Y, Sugiyama Y, Iga T, Hanano M (1986) Transport of cimetidine by the rat choroid plexus *in vitro*. *J Pharm Exp Ther* 239: 927–935

117   Suzuki H, Sawada Y, Sugiyama Y, Iga T, Hanano M (1987) Transport of benzyl penicillin by the rat choroid plexus *in vitro*. *J Pharm Exp Ther* 242: 660–665

118   Prichard JB, Sweet DH, Miller DS, Walden R (1999) Mechanism of organic anion transport across the apical membrane of the choroids plexus. *J Biol Chem* 274: 33382–33387

119   Ghersi-Egea J-F, Strazielle N (2002) Choroid plexus transporters for drugs and other xenobiotics. *J Drug Targ* 11: 353–357

120   Hosoyamada M, Sekine T, Kanai Y, Endou H (1999) Molecular cloning and functional expression of a multi specific organic anion transporter from human kidney. *Am J Physiol* 276: F122–F128

121   Kusuhara H, Sugiyama Y (2001) Efflux transport systems for drugs at the blood-brain and blood cerebrosinal fluid barrier (Part 1). *Drug Discov Today* 6: 150–156

122   Kusuhara H, Sugiyama Y (2001) Efflux transport systems for drugs at the blood-brain and blood cerebrosinal fluid barrier (Part 2). *Drug Discov Today* 6: 206–212

123   Begley DJ (2003) Efflux mechanisms in the CNS: A powerful influence on drug distrib-

ution within the brain. In: HS Sharma, J Westman (eds): *Blood-spinal cord and brain barriers in health and disease*. Elsevier/Academic Press, San Diego, 83–96

124 Meier PJ, Eckhardt U, Schoeder A, Hagenbuch, Steiger, B (1997) Substrate specificity of sinusoidal bile acid and organic anin uptake systems in rat and human liver. *Hepatology* 26: 1667–1677

125 Noé B, Hagenbuch B, Stieger B, Meier PJ (1997) Isolation of a multispecific organic anion and cardiac glycoside transporter from rat brain. *Proc Natl Acad Sci USA* 94: 10346–10350

126 Li L, Lee TK, Meier PJ, Ballatori N (1998) Identification of glutathione as a driving force and leukotriene C4 as a substrate for Oatp1, the hepatic sinusoidal organic solute transporter. *J Biol Chem* 273: 16184–16191

127 Reichel C, Gao B, van Monyfoort J, Cattori V, Rahner C, Hagenbuch B, Stieger B, Kamisako T, Meier PJ (1999) Localization and function of the organic anion transporting polypeptide Oatp2 in rat liver. *Gastroenterology* 117: 688–695

128 Kakyo M, Sakgami H, Nishio T, Naki D, Nakagomi R, Tokui T, Naitoh T, Matsuno S, Abe T, Yawo H (1999) Immunohistochemical distribution and functional characterisation of an organic anion transporting polypeptide 2 (Oatp2). *FEBS Lett* 445: 343–346

129 Gao B, Hagenbuch B, Kullak-Ublick GA, Benke D, Aguzzi A, and Meir PJ (2000) Organic anion transporting polypeptides mediate transport of opioid peptides across blood-brain barrier. *J Pharmacol Exp Ther* 294: 73–79

130 Kullak-Ublick GA, Ismair MG, Stieger B, Landmann L, Huber R, Pizzagalli F, Fattinger K, Meir PJ, Hagenbuch B (2001) Organic anion-transporting polypeptide B (OATP-B) and its functional comparison with three other OATPs of human liver. *Gastroenterology* 120: 525–533

131 Satlin LM, Amin V, Wolkoff AW (1997) Organic anion transporting polypeptide mediates organic anion/HCO_3^- exchange. *J Biol Chem* 272: 26340–26345

132 Angeletti RH, Novikoff PM, Juvadi SR, Fritschy JM, Meir PJ, Wolkoff AW (1997) The choroid plexus is the site of the organic anio transport protein in the brain. *Proc Natl Acad Sci USA* 94: 283–286

133 Gao B, Stieger B, Noé B, Fritschy J-M, Meir PJ (1999) Localization of the organic anion transporting polypeptide 2 (Oatp2) in capillary endothelium and choroid plexus epithelium of rat brain. *J Histochem Cytochem* 47: 1255–1263

134 Gao B, Meier PJ (2001) Organic anion transport across the choroid plexus. *Microsc Res Tech* 52: 60–64

135 Sweet DH, Miller DS, Pritchard JB (2001) Ventricular choline transport: a role for organic cation transporter 2 expressed in choroid plexus. *J Biol Chem* 276: 41611–41619.

136 Kekuda R, Prasad PD, Wu X, Wang H, Fei Y-J, Leibach, FH, Ganapathy V (1998) Cloning and functional chacterisation of a potential sensitive, polyspecific organic cation transporter (OCT3) most abundantly expressed in placenta. *J Biol Chem* 272: 15971–15979

137 Miller DS, Villalobos AR, Pritchard JB (1999) Organic cation transport in rat choroids plexus cells studied by fluorescent microscopy. *Am J Physiol* 45: C955–C968

138 Tamai I, Ohashi R, Nezu J, Yabuuchi H, Oku A, Simane M, Sai Y, Tsuji A (1998) Molecular and functional indentification of sodium ion-dependent, high affinity human carnitine transporter OCTN2. *J Biol Chem* 272: 20378–20382

139 Wu X, Huang W, Prasad PD, Seth P, Rajan DP, Leobach FH, Chen J, Conway SJ, Ganapathy V (1999) Functional characteristics and tissue distribution pattern of organic cation transporter 2 (OCTN2), an organic cation/carnitine transporter. *J Pharm Exp Ther* 290: 1482–1492

140   Wu X, George RL, Huang W, Wang H, Conway SJ, Leibach FH, Ganapathy V (2000) Structural and functional characteristics and tissue distribution of rat OCTN1 an organic cation transporter, cloned from placenta. *Biochim Biophys Acta* 1466: 315–327

141   Yabuuchi H, Tamai I, Nezu J, Sakamoto K, Oku A, Shimane M, Sai Y, Tsuji A (1999) Novel membrane transporter OCTN1 mediates multispecific, bi-directional and pH dependent transport of organic cations. *J Pharmacol Exp Ther* 289: 768–773

142   Kido Y, Tamai I, Ohnaraia A, Sai, Y, Kagami T, Nezu J, Nikaido H, Hashimoto N, Asano M, Tsuji A (2001) Functional relevance of carnitine transporter \octn2 to brain distribution of L-carnitine and acetyl-L- carnitine across the blood-brain barrier. *J Neurochem* 79: 959–969

143   Tamai I, Ohashi R, Nezu J, Sai, Y, Kobayashi, D, Oku A, Simane M, Tsuji, A (2000) Molecular and functional characterisation of organic cation/carnitine transporter family in mice. *J Biol Chem* 275: 40064–40072

144   Klien I, Sarkadi B, Váradi A (1999) An inventory of the human ABC proteins. *Biochim Biophys Acta* 1461: 237–262

145   Terasaki T, Hosoya K (1999) The blood-brain barrier efflux transporters as a detoxifying stsytem for the brain. *Adv Drug Deliv Rev* 5: 195–209

146   Sun H, Dai H, Shaik N, Elmquist WF (2003) Drug efflux transporters in the CNS. *Adv Drug Deliv Rev* 55: 83–105

147   Deely RG, Cole SP (1997) Function, evolution and structure of multidrug resistance protein (MRP). *Seminars in Cancer Biol* 8: 193–204

148   Chin JE, Soffir R, Noonan KE, Choi KY, Robinson IB (1989) Structure and expression of the human MDR (P-glycoprotein) gene family. *Mol Cell Biol* 9: 2808–3820

149   Hsu SIH, Cohen D, Kirchner LS, Hartstein M, Horwitz SB (1990) Structural analysis of the mouse mdr1a (P-glycprotien) promoter reveals the basis for differential transcripts heterogeneity in multidrug resistant J774.2 cells. *Mol Cell Biol* 10: 3596–3606

150   Cordon-Cardo C, O'Brien JP, Casals D, Rittman-Grauer L, Biedler JL, Melamed MR, Bertino JR (1989) Multidrug-resistance gene (P-glycoprotein) is expressed by endothelial cells at blood-brain barrier sites. *Proc Natl Acad Sci USA* 86: 695–698

151   Tatsuta T, Naito M, Oh-hara T, Sugawara I, Tsuruo T, (1992) Functional involvement of P-glycoprotein in blood-brain barrier. *J Biol Chem* 267: 20383–20391

152   Stewart PA, Beliveau R, Rogers KA (1996) Cellular localisation of P-glycoprotein recognised by MRK16 monoclonal antibody in brain and spinal cord. *J Histochem Cytochem* 44: 679–685

153   Beaulieu É, Demeule M, Ghitescu L, Béliveau R (1997) P-glycoprotein is strongly expressed in the luminal membranes of the endothelium of blood-vessels in the brain. *Biochem J* 326: 301–307

154   Begley DJ, Khan EU, Rollison C, Abbott NJ, Regina A, Roux F (2000) The role of brain extracellular fluid production and efflux mechanisms in drug transprt to the brain. In: DJ Begley, MWB Bradbury, J Kreuter (eds): *The blood-brain barrier and drug delivery to the CNS.* Marcel Dekker, New York, 93–108

155   Rao VV, Dahlheimer JL, Bardgett ME, Snyder AZ, Finch RA, Sartorelli AC, Piwnica-Worms D (1999) Choriod plexus epithelial expression of MDR1 P-gycoprotein and multidrug resistance-associated protein contribute to the blood-cerebrospinal fluid drug-permeability barrier. *Proc Natl Acad Sci USA* 96: 3900–3905

156   Regina A, Koman A, Piciotti M, El Hafny B, Center MS, Bergman R, Couraud P-O, Roux

F (1998) Mrp1 multidrug resistance-associated protein and P-glycoprotein expression in rat brain microvessel endothelial cells. *J Neurochem* 71: 705–715

157   Kusuhara H, Sugiyama Y (2002) Role of transporters in the tissue-selective distribution and elimination of drugs: Transporters in the liver, small intestine, brain and kidney. *J Control Rel* 78: 43–54

158   Miller DS, Nobmann SN, Gutmann H, Toeroek J, Drewe G, Fricker G (2000) Xenobiotic transport across isolated brain microvessels studied by confocal microscopy. *Mol Pharmacol* 58: 1357–1367

159   Fricker G, Nobmann S, Miller DS (2002) Permeability of porcine blood-brain barrier to somatostatin analogues. *Br J Pharmacol* 135: 1308–1314

160   Zhang Y, Han H, Elmquist WF, Miller DW (2000) Expression of various multidrug resistance-associated protein (MRP) homologues in brain mircovessel endothelial cells. *Brain Res* 876: 148–153

161   Cisternino S, Rouselle C, Lorico A, Rappa G, and Scherrmann J-M (2003) Apparent lack of MRP1-mediated efflux at the luminal side of mouse BBB endothelial cells. *Pharm Res* 20: 904–909

162   Wijnholds J, De Lange EL, Scheffer GL, van den Berg DJ, Mol CA, van der Walk M, Schinkel AH, Scheper RJ, Briemer DD, Borst P (2000) Multidrug resistance protein protects the choroid plexus epithelium and contributes to the blood-cerebrospinal fluid barrier. *J Clin Invest* 105: 279–285

163   Eisenblätter T, Galla H-J (2002) A new multidrug resistance protein at the blood-brain barrier. *Biochem Biophys Res Comm* 293: 1273–1278

164   Cooray H, Blackmore CG, Maskell L, Barrand MA (2002) Localisation of breast cancer resistance protein in microvessel endothelium of human brain. *NeuroReport* 13: 2059–2063.

165   Zhang W, Cui H, Zhang H, Ball M, Stanimirovich D (2002) Expression and functional characterisation of the ABC transporter gene, ABCG2, in brain endothelial cells. *Abstract: Human Genome Meeting 2002*. Nature Publishing Group, Shanghai, China, 122

166   Litman T, Brangi M, Hudson E, Fetsch P, Abati A, Ross DD, Miyake K, Resau, JH, Bates SE (2000) The multidrug-resistant phenotype associated with over expression of the new ABC half transporter, MXR (ABCG2). *J Cell Sci* 113: 2011–2021

167   Banks WA, Kastin A (1996) Passage of peptides across the blood-brain barrier: Pathophysiological perspectives. *Life Sci* 59: 1923–1943

168   Pan W, Banks WA, Kastin AJ (1997) Permeability of the blood-brain barrier and spinal cord barriers to interferons. *J Neuroimmunol* 76: 105–111

169   Pan W, Banks WA, Kastin AJ (1990) Permeability of the blood-brain barrier to neurotrophins. *Brain Res* 788: 87–94

170   Kastin AJ (1999) Multiple direct actions of peptides on the CNS. *Brain Res Bull* 50: 361–362

171   Kastin AJ, Pan W, Maness LM, Banks WA (1999) Peptides crossing the blood-brain barrier: some unusual observations. *Brain Res* 848: 96–100

172   Teuscher NS, Novotny A, Keep R.F, Smith DE (2000) Functional evidence for presence of PEPT2 in rat choroid plexus: studies with glycylsarcosine. *J Pharm Exp Ther* 294: 494–499

173   Yamashita T, Shimada S, Guo W, Sato K, Kohmura E, Hayakawa T, Tahagi T, Tohyama M (1997) Cloning and functional expression of a brain peptide/histidine transporter. *J Biol Chem* 272: 10205–10211

174   Minn A, El-Bachá DS, Bayol-Denizot C, Lagrange P, Suleman FG (2000) Drug metabo-

lism in the brain: benefits and risks. In: DJ Begley, MWB Bradbury, J Kreuter (eds): *The blood-brain barrier and drug delivery to the CNS*. Marcel Dekker, New York, 145–170

175   Minn A, Leclerc S, Heydel J-M, Minn A-L, Denizot C, Cattarelli M, Netter P, Gradinaru D (2002) Drug transport into the mammalian brain: the nasal pathway and its specific metabolic barrier. *J Drug Targ* 10: 285–296

176   Ghersi-Egea J-F, Strazielli N (2001) Brain drug delivery, drug metabolism, and multidrug resistance at the choroid plexus. *Microsc Res Tech* 52: 83–88

177   Rush and Hersh LB (1982) Multiple molecular forms of rat brain enkephalinase. *Life Sci* 21: 445–451

178   Turner AJ (1987) Endopeptidase-24.11 and neuropeptide metabolism. In: AJ Turner (ed): *Neuropeptides and their peptidases*. Ellis Horwood and UCH Verlagsgesellschaft, Chichester, UK, 310–315

179   Skidgel, RA, Defendini R, Erdos EG (1987) Angiotensin I converting enzyme and its role in neuropeptide metabolism. In: AJ Turner (ed): *Neuropeptides and their peptidases*. Ellis Horwood and UCH Verlagsgesellschaft, Chichester, UK, 165–182

180   Dyer SH, Slaughter CA, Orth, K, Moomaw, CR, Hersh LB (1990) Comparison of the soluble and membrane bound forms of the puromycin-sensitive enkephalin-degrading aminopeptidase from rat. *J Neurochem* 44: 1427–1435

181   Brownlees J, Williams CH (1993) Peptidases, peptides and the mammalian blood-brain barrier. *J Neurochem* 60: 793–803

182   Witt KA, Gillespie TJ, Huber JD, Egleton RD, Davis TP (2001) Peptide drug modifications to enhance bioavailability and blood brain barrier permeability. *Peptides* 22: 2329–2343

Progress in Drug Research, Vol. 61
(L. Prokai and K. Prokai-Tatrai, Eds.)
©2003 Birkhäuser Verlag, Basel (Switzerland)

Peptide transport across the blood-brain barrier

By Abba J. Kastin and Weihong Pan

VA Medical Center and Tulane University School of Medicine
New Orleans, LO 70112-1262
USA

Abba J. Kastin

was born in Ohio and educated at Harvard College and then Harvard Medical School. After starting his work on neuropeptides at NIH, he moved to New Orleans where he still lives. Dr. Kastin has been made an honorary member of seven medical societies outside the United States, has received two honorary doctorates (one in USA and one foreign), and has won several national and international awards. Author of 800 papers, he has been listed among the 100 researchers most cited in the scientific literature. He also is Editor-in-Chief of Peptides: an International Journal *and President of the International Neuropeptide Society.*

Weihong Pan

M.D. (1990, Shanghai Medical University), Ph.D. (1997, Tulane University Neuroscience Program), is a clinical neurologist certified by the American Board of Psychiatry and Neurology, and an Associate Professor in Medicine at Tulane University. Dr. Pan's research interest is mainly the regulation of cytokine transport at the blood-brain and blood-spinal cord barrier. This includes identification of transport systems, intracellular trafficking, and implications of transport in spinal cord injury and other CNS pathology. Dr. Pan has published more than forty full papers and five book chapters, and has been an invited speaker at several international meetings. She is on the editorial board of Peptides *and* Current Pharmaceutical Design.

Contents

Key words

Blood-brain barrier, peptides, feeding.

Glossary of abbreviations

BBB, blood-brain barrier; BUI, brain uptake index; CSF, cerebrospinal fluid; CVOs, circumventricular organs; HPLC, high performance liquid chromatography; iv, intravenous; IL, interleukin.

1   Introduction

No longer considered a static, impenetrable barrier, the dynamic regulatory functions of the blood-brain barrier (BBB) have become increasingly apparent. This is particularly evident for the transport of peptides across the BBB. Interactions of peptides with the endothelial cells composing the BBB could lead to endocytosis and eventually transcytosis of peptides. It is helpful to understand pertinent characteristics of the BBB, the circumventricular organs (CVOs), and choroid plexuses. Here we describe the methods used for measurement of the passage of peptides across the BBB, illustrate the types of interactions of peptides with the BBB, and summarize current results with the BBB penetration of feeding-related peptides.

Part of the previous skepticism concerning the ability of peptides to cross the BBB probably could be explained by the lack of appropriate and sensitive methods for quantification of the relatively tiny amount in blood that reaches the brain parenchyma in intact form. For insulin, only about 0.05% of the amount injected reaches the brain [1] and this low amount cannot be explained by rapid transport out of the brain [2]. For cyclo(His-Pro), even though only about 0.02% reaches the brain after intravenous (iv) administration, this amount is sufficient to reduce sleep time resulting from ethanol-induced narcosis [3].

For opiates, it was shown many years ago that peripheral injection of an analog of Met-enkephalin causes electroencephalogram (EEG) changes that can be reversed by an opiate antagonist that crosses the BBB, but not by one which only acts peripherally [4]. This early study further established that even the small amounts of peptide crossing the BBB can exert actions there. Although morphine is not a peptide, it is often misleadingly thought that most of it freely enters the brain; yet only about 0.02% of the injected dose is found there [5, 6] and only 0.002% of the injected shellfish toxin domoic acid [7], although other factors such as efflux may be involved.

In general, only a small portion of peripherally delivered peptide reaches brain parenchyma. However, the amount is often sufficient to affect physiological functions and modulate neuroendocrine and behavioral responses. Most of the effects involve direct passage across the BBB, as will be further discussed below. Thus, the study of where and how interactions take place at the BBB will provide essential information that could be used to modify the neurovascular interface and help design peptide therapeutics.

2 The relative roles of the BBB, CVOs, and choroid plexuses in the passage of peptides from blood to brain

The BBB is formed by cerebral capillaries devoid of the gaps between their endothelial cells that occur elsewhere in the body. There are few fenestrations, pinocytic vesicles, or channels providing easy access of peripheral peptides to the brain. Instead, these capillary endothelial cells fit together with tight junctions lined by a continuous basement membrane on the brain side [8].

Capillaries in the CVOs do not have the typical features of the BBB and therefore appear to provide more direct communication for peptides in the peripheral blood. The peptides, however, cannot penetrate beyond the CVOs to the rest of the brain within the BBB. This restriction is explained by tight junctions between ependymal cells which constitute a border between the CVOs and the rest of the brain [9].

The small capacity of the CVOs to transport peptides from blood to brain can be put into perspective by consideration of surface areas. The total surface area of the CVOs is only 0.02 cm^2/g brain tissue whereas that of the BBB is 100 to 150 cm^2/g of the BBB; this represents a 5000- to 7500-fold difference [10, 11]. Therefore, when radioactively labeled interleukin (IL)-1α is delivered into blood, less than 5% of the total radioactivity representing intact IL-1α in brain 20 min later is contributed by the CVOs [12, 13].

The choroid plexuses line the ventricles of the brain. Like the CVOs, the endothelial cells of their capillaries are also devoid of tight junctions, but the apical part of their epithelial cells lining the cerebrospinal fluid (CSF) are sealed by tight junctions [9]. Some reports show that leptin enters the brain through the choroid plexus and median eminence (which is a CVO) [14–16], but we found no significant difference between the entry of ^{3}H-leptin or ^{125}I-leptin into the cortical area of the brain (which does not contain choroid plexuses and CVOs) and entry into the remainder of the brain [17]. Furthermore, Flier and his associates do not feel that the choroid plexus and CSF is a major route for leptin reaching the hypothalamus [18]. They indicate that the low CSF leptin concentrations are not sufficient to activate Janus kinase-STAT signaling.

3   Technical considerations in quantifying BBB passage of peptides

Since most peptides and polypeptides we study are large hydrophilic molecules that have slow and limited penetration across the BBB, traditional techniques such as the BUI (brain uptake index) method [19] are not appropriate and have led to the misconception that "circulating peptides do not effectively cross the BBB" and that "it is unlikely that distribution of peptides in brain will occur after systemic administration" [20]. The BUI method measures single-pass uptake from blood over a few seconds and provides an excellent estimate of the rapid uptake of glucose and water by the brain. "However, as so often stated, the BUI technique is unsuitable for study of slowly penetrating molecules" [8].

There are three basic methods to evaluate how much and how fast peptides cross the BBB: iv delivery followed by multiple-time regression analysis [17, 21–23], *in situ* brain perfusion [24–26], and studies in cultured brain microvessel endothelial cells constituting the *in vitro* BBB [27–30]. For each, the prerequisite is that the peptide remain stable during the study period and that it completely cross the endothelial cell barrier.

3.1   Determination of the integrity and degradation patterns of the peptide in blood and brain by high performance liquid chromatography (HPLC)

No matter what technique is used to determine that a peptide injected peripherally reaches the brain, it must be established that the peptide remains intact. Otherwise, a fragment of the injected substance could mistakenly appear to represent the intact peptide as measured by immunoreactivity or radioactivity.

Chromatographic measurement, especially HPLC, is a convenient way to ensure that the injected peptide reaches the brain in intact form and to easily quantify the percentage and pattern of degradation. For large peptides and polypeptides, use of SDS-PAGE (polyacrylamide gel electrophoresis) also can be helpful. Both of these chromatographic tools will identify intact peptide, any metabolic products, and the free radioactive label.

Of the peptides we have studied, most remain relatively stable in blood for at least 10 minutes after injection into the blood stream. However, we have preliminary data suggesting that 10 minutes after iv injection some peptides radioactively labeled with ^{125}I are more than 50% degraded in blood; these include orexin B, galanin, peptide YY (PYY 1-36), urocortin III (stresscopin), neuropeptide W-23, [6-Br-L-Trp]-neuropeptide B-29, and the polypeptide IL-11. Oncostatin M, NEP(1-40), and PYY(3-36) remained between 50 - 60% intact after 10 min. Some others, like prolactin-releasing peptide-31 (PrP), are rapidly degraded by brain, even after allowance for the amount of metabolism occurring during the process of homogenization of brain tissue with its consequent liberation of intracellular enzymes.

Degradation can occur after passage across the BBB and entry into brain before the homogenization process. If this happens, the reported values for intact peptide reaching the brain would be underestimated. In addition, some degradation can occur at the BBB itself resulting from peptidases present in the capillary endothelial cells [31], and this can vary in microvessels that are obtained from different regions of the brain [32].

3.2 Estimation of compartmental distribution by capillary depletion

Even if HPLC indicates that the injected peptide has remained intact, this can still be misleading because the brain tissue being examined includes blood vessels. The radioactively labeled peptide that was injected might be associated with the brain vasculature but not the brain parenchyma. Therefore it is necessary to determine that the injected peptide has reached the brain parenchyma, which includes neurons, glia, and extracellular matrix. For example, although the apparent influx rate was comparable for the peptides epidermal growth factor (EGF) and transforming growth factor α (TGF-α), EGF enters the parenchyma of the brain whereas TGF-α is mainly associated with the cerebral vasculature without crossing the BBB [33, 34].

The capillary depletion method [35] takes advantage of the different density of microvessels and brain parenchyma. It involves dextran density centrifugation which results in the cerebral microvessel endothelial cells forming a pellet while the parenchyma remains in the supernatant. Effective separation of the vessels can be verified by high concentrations of γ-glutamyl transpeptidase and alkaline phospatase in the pellet and minimal concentrations in the supernatant [35, 36]. A further correction is applied by subtraction of the vascular space as determined from the volume of distribution of the ^{99m}Tc-albumin co-injected with the peptide. A washout procedure involving intracardioventricular perfusion removes the loosely adherent material in the lumen of the vasculature but not that endocytosed or tightly bound to the wall of the capillary endothelial cells. Furthermore, acid wash can help to differentiate peptide that is bound to the transporter or other binding sites and that which has been internalized within the endothelial cells. Thus, the amount of injected radiolabeled peptide found in several compartments can be determined and this must be done to establish that the injected peptide has reached brain parenchyma.

3.3 Estimation of influx rate and volume of distribution by multiple-time regression analysis

Multiple-time regression analysis [21, 22, 37] is a sensitive and reliable measurement of the influx rate and volume of distribution of a peptide that is relatively stable in blood and has slow penetration across the BBB. The peptide, usually radioactively labeled, is delivered iv or intra-arterially as a bolus. At various times after the initial injection, blood and brain are sampled after decapitation of the anesthetized animal, and the amount of peptide in each compartment is estimated by measurement of the radioactivity in both blood (cpm/µl of serum) and brain (cpm/g of tissue). The brain/blood ratio of radioactivity at that particular study time is then calculated.

For most small peptides, the study time is about 10 min; during this interval the peptide remains intact after iv injection, as determined by HPLC. Sampling at each minute, for example, provides sufficient data for regression correlation analysis between the brain/blood ratio of radioactivity and the time of iv circulation of the peptide. Since each peptide has a different disappearance curve in blood, and its half-life is also affected by the addition of modulators of its potential transport depending on the study design, a standardized "exposure time" is used instead of real time.

Exposure time is the theoretical steady-state value at that particular study time if the blood concentration of the peptide remained constant. Exposure time at time t is calculated as the integral of serum radioactivity over time divided by the radioactivity at time t. The regression correlation between the brain/blood ratio of peptide radioactivity and exposure time is linear. The slope of this regression line represents the rate of influx (K_i) of the peptide from blood to brain, and the y-intercept at time 0 is the initial volume of distribution in the brain.

The sensitivity of the measurement of trace amounts of radioactivity of injected peptide in the brain is higher than with other methods such as microdialysis, immunohistochemistry or bioassay, and its quantification is much more direct. An additional advantage is that the radioactive tracer method differentiates the exogenous peptide being measured from any confounding endogenous peptide, which can be difficult to determine by alternative techniques.

One of the major applications of multiple-time regression analysis is to establish whether the peptide enters the brain by a saturable transport system. This is indicated when co-administration of excess unlabeled peptide together with the labeled peptide results in a significant inhibition of the rate of influx.

3.4   Determination of vascular diffusion

As a control, it is necessary to determine whether non-specific entry occurs during the time used for measurement of the K_i by multiple-time regression analysis. It might seem that injection of the denatured form of the peptide being tested would provide the best control, but substantial changes in conformation of the peptide can occur after denaturation [17]. It also might seem that other inert compounds with molecular weights similar to that of the peptide being studied would be suitable, but it is difficult to account for possible differences in physicochemical properties such as lipophilicity, hydrogen bonding and electrical charge.

An additional problem occurs for a marker like inulin in that it is not commercially available in a form that can be measured in a gamma counter [17]; the required liquid scintillation (beta) counting is much more cumbersome. It might also seem desirable to inject the peptide labeled by a different radioactive tracer in the period soon after completion of the experiment followed by subtraction of the uptake of the second tracer, but this can underestimate the entry rate of a peptide with relatively fast uptake [38].

The simplest control involves co-injection of a vascular marker such as albumin together with the peptide (^{125}I-peptide + ^{99m}Tc-albumin) [37]. Albumin does not enter the brain during the usual time period used for determination of the influx rate of peptides. It has the added advantage of ensuring that the BBB remains intact during the time in which the entry of the peptide is being measured.

3.5 Passage of peptides across the *ex vivo* BBB: *in situ* brain perfusion

Studies of peptides interacting with the BBB in intact animals are not always easy. For determinations in blood, one must take multiple factors into consideration including tissue distribution, hepatic metabolism, renal excretion, peptidases and nonspecific degradation in blood, protein binding and self-aggregation, endogenous production and stimulation of acute phase reactions. Therefore, delivery of radioactively labeled peptides at a constant rate in blood-free perfusion buffer [24] obviates most worries, especially enzymatic degradation and protein binding in blood.

Perfusion methods vary depending on the species of the animal used and have been described in detail elsewhere [24, 38]. Some limitations are that: (a) this method is less physiological than when the peptide reaches the brain as part of the normal circulation, (b) perfusion time is usually short despite oxygenation of perfusion buffer and use of erythrocytes as the oxygen carrier, since the integrity of the BBB is eventually compromised after sustained ischemia and hypoxia and (c) perfusion rate, temperature, duration and components of the perfusion buffer affect the interpretation and comparison of results from different experimenters. Nonetheless, *in situ* brain perfusion provides an easily modifiable milieu for study of the direct interactions of peptides with the BBB, and interpretation of the results is usually straightforward, especially when a paracellular permeability marker such as albumin is used simultaneously.

3.6 Peptide passage in the opposite direction: brain-to-blood efflux

If a radioactively labeled peptide is transported out of the brain soon after it is injected, measurement of its counts in the brain will be misleadingly low, or even absent. Since albumin is removed from brain by the bulk flow of CSF, comparison of its rate of disappearance with that of the peptide being studied will indicate whether a transport system exists for that peptide [39, 40]. Greater certainty can be achieved by determination of any inhibition of efflux after intracerebroventricular co-injection of excess unlabeled peptide. There is a high degree of correlation for several compounds between their usually determined rate of disappearance from brain and their rate of appearance in blood [41].

The p-glycoprotein system transports many compounds out of the brain, effectively preventing accumulation of several toxic substances in the brain, but also serving as an obstacle to delivery of therapeutic agents in other instances [42–44]. Substrate specificity seems to be confusing for opiate peptides, in that endorphin [45] but not Met-enkephalin, Tyr-MIF-1, endomorphin-1 or endomorphin-2 [46] are transported out of the brain by this system.

3.7 *In vitro* models

Although brain microvessel endothelial cells in culture are never exactly the same as those in the intact body in terms of permeability features, isolated cells make convenient models for study of the mechanisms of transport of peptides. Transport or uptake studies can be performed in cells grown on a solid support phase, yet most models use transwell diffusion systems. The acceptor and donor chambers can be assembled side-by-side or up and down. The latter vertical system has gained more popularity in recent years because of its ease of assembly and development of co-culture with astroglia which reinforce barrier function. A more complicated model involving growth of the endothelial cells in a complex tubular system has not been used widely and probably is not efficient for pharmacokinetic studies.

A successful *in vitro* BBB system should demonstrate relatively high transmembrane electrical resistance and low permeability to paracellular markers. Whether it is a purified primary brain microvessel endothelial cell culture or an immortalized endothelial cell line, it must show barrier-specific features such as expression of von Willebrand factor, abundance of alkaline phosphatase and γ-glutamyl transpeptidase and establishment of tight junctions. Once these requirements are met and appropriate controls are established, apical-to-basolateral flux (reflecting blood to brain penetration) and basolateral-to-apical flux (brain to blood) can be quantified.

The transwell system is convenient for determination of the intracellular degradation of peptides. After passage across the monolayer of endothelial cells, the radioactively labeled peptide is analyzed by HPLC, gel electrophoresis or acid precipitation. For instance, the radioactivity recovered from the basolateral side of the transwell 30 min after application of ^{125}I-EGF to the apical chamber containing TM-BBB4 endothelial cells represents only free iodine (Pan et al., unpublished observation).

Peptides could be internalized in the endothelial cells by fluid-phase endocytosis, by adsorptive-mediated endocytosis, or by receptor-mediated endocytosis. The last two are saturable processes. The binding receptor may not necessarily be the transporter. One example is basic fibroblast growth factor (bFGF), which uses adsorptive-mediated endocytosis [47] as does the HIV-1 protein gp120 [48, 49]. Similarly, although the opiate peptides Tyr-MIF-1 and Met-enkephalin share a transport system, the transporter is not either of their receptors [50].

The next tricky question is when and how does endocytosis lead to transport? Endocytosis is not equivalent to transcytosis, as most internalized ligands, regardless of mediation by clathrin-coated pits, caveolae, or other mechanisms, are usually directed to degradation pathways in the cellular vesicles. For leptin, this is illustrated by experiments involving binding and immunofluorescent microscopy in which internalization and subcellular localization of receptor isoforms occur without any evidence for special intracellular trafficking indicative of transcytotic transport [51, 52].

Methods to study endocytosis and intracellular trafficking in the brain microvessel endothelial cells are not novel since research on epithelial cell transport and immune cell antigen processing has been far more advanced than in the BBB area. Improved imaging techniques with electron and fluorescence micrographs in recent years have made intracellular localization and tracing easier for peptides. Site-directed mutagenesis and the yeast two-hybrid system also are powerful tools.

4 Mechanisms of peptide-BBB interactions

4.1 Saturable transport

The existence of saturable transport systems for peptides well illustrates the dynamic nature of the BBB. The BBB can limit the amount of peptide entering the brain if its concentrations in blood are high. This can be shown experimentally by co-injection of an excess of the unlabeled peptide. Inhibition of entry would not be seen if the peptide entered the brain from the blood through a leaky BBB or by simple passive diffusion.

The mechanisms responsible for saturable transport have not been well delineated. Although it appears that receptor mediated endocytosis is involved to some extent in the influx of insulin [53], TNF-α [54], and leptin [55], there is little evidence that this occurs with most other peptides. For the few peptides examined, this mechanism does not seem to be involved in the transport of EGF in the apical-to-basolateral direction [34] or, as mentioned above, the saturable transport system shared by Tyr-MIF-1 and Met-enkephalin [50].

4.2 Passive diffusion

Many peptides cross from blood to brain by passive diffusion based on physicochemical properties like lipophilicity and hydrogen bonding. Size also may be a limiting factor, and the rate of diffusion is inversely related to the square root of the molecular weight [56, 57]. As discussed above, co-injection of a vascular marker like albumin ensures that this cannot be explained by leakage between cells but occurs by passage through the cell.

4.3 Capture by the cerebral vasculature

Determination of the extent to which a peripherally injected peptide reaches the brain usually involves measurement of radioactivity or immunoreactivity in brain tissue. This can be misleading if the peptide remains trapped in the vasculature within the brain without reaching the parenchyma.

Some of the trapped peptide is loosely adherent and can be easily removed by a washout procedure after perfusion, as found with adrenomedullin (ADM) [58] and glucagon-like peptide (GLP)-1 [59]. Alternatively, some of the trapped peptide could be tightly bound to the lumen of the endothelial cells of the cerebral capillaries, as illustrated by TGF-α [33].

4.4 Protein binding and aggregation

A peripherally injected peptide might not cross the BBB to enter the brain if it became bound to a large protein in blood or self-aggregated. Protein binding is well known in blood, but aggregation of peptides is not, although it was described in 1984 [60].

During investigation of the crossing of the BBB, such phenomena are suspected when the peptide does not enter brain after iv administration but does enter after perfusion in blood-free buffer. Using SDS-polyacrylamide gel electrophoresis and capillary zone electrophoresis (CZE), we were able to determine that agouti-related protein (AgRP) (83-132) aggregates [61] whereas melanin-concentrating hormone (MCH) shows protein binding as well as self-aggregation [62], both interfering with ready penetration of the BBB. It is possible, however, that protein binding could serve as a degradation-free carrier in blood, conveying a peptide to the active surface of the BBB; this may happen with galanin-like peptide (GALP) [63].

5 Illustrative examples with peptides affecting food intake

Most of the issues discussed up to this point in the review are well illustrated by peptides involved in food ingestion. The remainder of this review mentions some unusual aspects of their interaction with the BBB, the peptides being presented in alphabetical order.

- *ADM*: The most unusual aspect of the interaction of ADM with the BBB was detected by the capillary depletion procedure. Although intact ADM appeared to reach the brain after peripheral administration, much of it remained loosely adherent in the lumen of the capillaries, easily washed out by perfusion [58]. This may suggest an action of ADM at the cerebral vascular level.
- *AgRP*: AgRP is unusual in that it self-aggregates as a trimer [61]. This probably explains its poor penetration of the BBB.
- *Amylin and pancreatic polypeptide*: Even though both amylin and pancreatic polypeptide (PP) are synthesized in the pancreas, only PP has a saturable transport mechanism for entering the brain [64, 65].
- *Cocaine and amphetamine-regulated transcript (CART)*: CART crosses the BBB at a rapid rate which might imply the existence of a saturable transport system. However, this is not the case; it crosses by passive diffusion not readily explained by lipophilicity [65a].
- *Corticotropin-releasing hormone (CRH)*: CRH is somewhat unusual in that it has a saturable transport system out of the brain [66]. This efflux is energy dependent, but p-glycoprotein independent [67]. It is vigorous enough for the CRH to exert peripheral effects [68]. By contrast, the saturable influx system for CRH is relatively slow [69].

- *Cyclo(His-Pro)*: This endogenous TRH-related dipeptide does not cross the BBB by a saturable mechanism. Nevertheless, peripheral injection of cyclo (His-Pro) is sufficient to reverse the narcosis induced by ethanol, probably related to its high lipophilicity and resistance to enzymatic degradation [3].
- *GALP*: The blood concentrations of this galanin-related peptide are reduced by fasting [63]. Its entry into brain is also reduced by fasting. Not yet reported at the time of writing this review, these results suggest that unlike the stimulatory effect of galanin on food intake [70], GALP probably has a satiety effect.
- *Insulin*: Partial inhibition of the saturable blood-to-brain transport of insulin [53] occurs at concentrations of insulin that are found in the normal circulation [71, 72]. When production of insulin is eliminated by pretreatment with streptozotocin, brain entry of insulin increases [73].
- *Leptin*: Among the endogenous substances affecting feeding, most attention has been given to leptin. Leptin is mainly synthesized in fat cells [74] but is saturably transported across the BBB in order to exert its satiety effect [14]. Despite this transport system, increased levels of leptin in blood are not sufficient to inhibit food ingestion although injection of leptin into the brain can do so. This has led to the concept of "leptin resistance" which must involve the BBB [75–79]. Although the short form of the leptin receptor explains much of the transport of leptin, an additional mechanism of entry is involved [55, 80]. What is particularly unusual is that in obese *ob/ob* mice which never produce any leptin, brain transport of leptin is unchanged [81].
- *Mahogany (1377-1428)*: Mahogany(1377-1428) enters the brain by a saturable transport system [82]. There is no cross-inhibition with AgRP even though the homozygous mahogany locus blocks the obesity resulting from ectopic expression of the agouti locus in brain [83].
- *MCH and melanocyte-stimulating hormone (MSH)*: Neither MCH nor MSH cross the BBB by saturable transport mechanisms [62, 84, 85]. Unlike leptin which is produced peripherally and exerts its satiety effect centrally, MSH exerts part of its satiety effect peripherally by lipid mobilization [86].
- *Neuropeptide Y (NPY)*: The blood-to-brain penetration of the BBB by NPY is non-saturable and somewhat slow, but it is significantly faster than that of the albumin control [87]. Its lipophilicity, as determined by the octanol:buffer partition coefficient, is low. Although passive diffusion across the BBB by peptides like NPY is based on physicochemical properties, of which

lipophilicity is generally considered important [88], lipophilicity does not easily explain the brain entry of NPY.

- *Orexins*: Both orexin A and orexin B enhance feeding behavior, but orexin B is so rapidly degraded in blood that it was not possible to quantify its rate of entry into brain. Orexin A is much more lipophilic than orexin B, and this helps its crossing of the BBB by diffusion [89].

- *Urocortin (UCN)*: Under normal conditions, UCN I is unusual in that it does not cross the BBB any faster than the vascular control. It is even more unusual in that its latent saturable transport system is activated by co-administration of leptin and pretreatment with glucose [90, 91]. This may involve a novel binding of UCN I to leptin. UCN II, by contrast, enters the brain by itself but does so by passive diffusion that is not affected by co-injection of leptin [69]. UCN III (stresscopin) is rapidly degraded at the BBB, with little reaching the brain in intact form [69].

6   Conclusion and future work

Understanding the mechanisms of how peptides interact with the BBB is our major approach toward revealing how the brain communicates with the rest of the body. Peptides in the periphery may act on the brain microvessel endothelial cells by binding to cell surface receptors or by transcytosis with subsequent actions on brain parenchyma. Pharmacokinetic studies involving radioactive tracers have proved that intact peptides cross the BBB. Quantification of passage shows that peptides may gain meaningful access to CNS tissue by simple diffusion or by saturable transport systems at the BBB.

The impact of transport across the BBB lies in the recognition that the BBB is an immense interface between the CNS and the rest of the body. It shows regional differences in permeability and specific uptake of peptides and proteins [1, 64, 80, 92, 93]. Thus, one could imagine that the sites of action for a feeding-related peptide are significantly greater than what would have occurred through vagal nerve reflexes, a neuroendocrine axis, retrograde transport, or direct action at the CVOs. Further studies are ongoing by our group and others concerning the morphology of peptide transport, involvement of the structures surrounding the BBB (the extracellular matrix, neuronal and glial components, etc.), regulation in pathological conditions and detailed membrane events and intracellular trafficking that lead to transcytosis. Eventually, we hope to be able to modulate and take advantage of the transport systems at the BBB to deliver peptides as therapeutic agents.

Acknowledgements

Supported by the Department of Veterans Affairs and NIH (DK 54880).

References

1   Banks WA, Kastin AJ (1998) Differential permeability of the blood-brain barrier to two pancreatic peptides: insulin and amylin. *Peptides* 19: 883–889

2   Cashion MF, Banks WA, Kastin AJ (1996) Sequestration of centrally administered insulin by the brain: effects of starvation, aluminum, and TNFα. *Horm Behav* 30: 280–286

3   Banks WA, Kastin AJ, Akerstrom V, Jaspan JB (1993) Radioactively iodinated cyclo(His-Pro) crosses the blood-brain barrier and reverses ethanol-induced narcosis. *Am J Physiol* 267: E723–E729

4   Kastin AJ, Pearson MA, Banks WA (1991) EEG evidence that morphine and an enkephalin analog cross the blood-brain barrier. *Pharmacol Biochem Behav* 40: 771–774

5   Banks WA, Kastin AJ (1994) Opposite direction of transport across the blood-brain barrier for Tyr-MIF-1 and MIF-1: comparison with morphine. *Peptides* 15: 23–29

6   Advokat C, Gulati A (1991) Spinal transection reduced both spinal antinociception and CNS concentration of systemically administered morphine in rats. *Brain Res* 555: 251–258

7   Preston E, Hynie I (1991) Transfer constants for blood-brain barrier permeation of the neuroexcitatory shellfish toxin, domoic acid. *Can J Neurol Sci* 18: 39–44

8   Davson H, Segal MB (1995) *Physiology of the CSF and Blood-Brain Barriers*. CRC Press, New York

9   Johanson CE (1995) Ventricles and cerebrospinal fluid. In: Conn PM (ed): *Neuroscience in Medicine*. Lippincott, Philadelphia, 171–196

10   Strand FL (1999) *Neuropeptides: Regulators of Physiological Processes*. MIT Press, Cambridge, MA, 89

11   Begley DJ (1994) Peptides and the blood-brain barrier: the status of our understanding. *Ann NY Acad Sci* 739: 89–100

12   Plotkin SR, Banks WA, Kastin AJ (1996) Comparison of saturable transport and extracellular pathways in the passage of interleukin-1α across the blood-brain barrier. *J Neuroimmunol* 67: 41–47

13   Maness LM, Kastin AJ, Banks WA (1998) Relative contributions of a CVO and the microvascular bed to delivery of blood-born IL-1α to the brain. *Am J Physiol* 275: E207–E212

14   Banks WA, Kastin AJ, Huang W, Jaspan JB, Maness LM (1996) Leptin enters the brain by a saturable system independent of insulin. *Peptides* 17: 305–311

15   Zlokovic BV, Jovanovic S, Miao W, Samara S, Verma S, Farrell CL (2000) Differential regulation of leptin transport by the choroid plexus and blood-brain barrier and high affin-

ity transport systems for entry into hypothalamus and across the blood-cerebrospinal fluid barrier. *Endocrinology* 141: 1434–1441

16   Thomas SA, Preston JE, Wilson MR, Farrell CL, Segal MB (2001) Leptin transport at the blood-cerebrospinal fluid barrier using the perfused sheep choroid plexus model. *Brain Res* 895: 283–290

17   Kastin AJ, Akerstrom V, Pan W (2001) Validity of multiple-time regression analysis in measurement of tritiated and iodinated leptin crossing the blood-brain barrier: meaningful controls. *Peptides* 22: 2127–2136

18   Bjorbaek C, Elmquist JK, Michl P, Ahima RS, van Bueren A, McCall AL, Flier JS (1998) Expression of leptin receptor isoforms in rat brain microvessels. *Endocrinology* 139: 3485–3491

19   Oldendorf W (1971) Brain uptake of radiolabeled amino acids, amines, and hexoses after arterial injection. *Am J Physiol* 221: 16929–1639

20   Pardidge WM (1984) Transport of nutrients and hormones through the blood-brain barrier. *Fed Proc* 43: 201

21   Blasberg RG, Fenstermacher JD, Patlak CS (1983) Transport of α-aminoisobutyric acid across brain capillary and cellular membranes. *J Cereb Blood Flow Metab* 3: 8–22

22   Patlak CS, Blasberg RG, Fenstermacher JD (1983) Graphical evaluation of blood-to-brain transfer constants from multiple time uptake data. *J Cereb Blood Flow Metab* 3: 1–7

23   Blasberg RG, Gazendam J, Patlak CS, Fenstermacher JD (1980) Quantitative autoradiographic studies of brain edema and a comparison of multi-isotope autoradiographic techniques. In: Cervos-Navarro J, Feuerstein GZ (eds): *Advances in Neurology*. Raven Press, New York, 255–270

24   Zloković BV, Lipovac MN, Begley DJ, Davson H, Rakíc L (1987) Transport of leucine-enkephalin across the blood-brain barrier in the perfused guinea pig brain. *J Neurochem* 49: 310–315

25   Zlokovic BV, Begley DJ, Djuricic BM, Mitrovic DM (1986) Measurement of solute transport across the blood-brain barrier in perfused guinea-pig brain: method and application to N-methyl-α-aminoisobutyric acid. *J Neurochem* 46: 1444–1451

26   Zlokovic BV, Mackic JB, Djuricic BM, Davson H (1989) Kinetic analysis of leucine-enkephalin cellular uptake at the luminal side of the blood-brain barrier of an *in situ* perfused guinea pig brain. *J Neurochem* 53: 1333–1340

27   Abbott NJ, Hughes CCW, Revest PA, Greenwood J (1992) Development and characterisation of a rat brain capillary endothelial culture: towards an *in vitro* blood-brain barrier. *J Cell Sci* 103: 23–37

28   Audus KL, Borchardt RT (1987) Bovine brain microvessel endothelial cell monolayers as a model system for the blood-brain barrier. *Ann NY Acad Sci* 507: 9–18

29   Terasaki T, Hosoya K. (2001) Conditionally immortalized cell lines as a new *in vitro* model for the study of barrier functions. *Biol Pharm Bull* 24: 111–118

30   Maresh GA, Maness LM, Zadina JE, Kastin AJ (2001) *In vitro* demonstration of a saturable transport system for leptin across the blood-brain barrier. *Life Sci* 69: 67–73

31   Egleton RD, Davis TP (1997) Bioavailability and transport of peptides and peptide drugs into the brain. *Peptides* 18: 1431–1439

32   Kastin AJ, Hahn K, Zadina JE (2001) Regional differences in peptide degradation by rat cerebral microvessels: a novel regulatory mechanism for communication between blood and brain. *Life Sci* 69: 1305–1312

33   Pan W, Vallance KL, Kastin AJ (1999) TGF alpha and the blood-brain barrier: accumulation in cerebral vasculature. *Exp Neurol* 454–459

34   Pan W, Kastin AJ (1999) Entry of EGF into brain is rapid and saturable. *Peptides* 20: 1091–1098

35   Triguero D, Buciak J, Pardridge WM (1990) Capillary depletion method for quantification of blood-brain barrier transport of circulating peptides and plasma proteins. *J Neurochem* 54: 1882–1888

36   Gutierrez EG, Banks WA, Kastin AJ (1993) Murine tumor necrosis factor alpha is transported from blood to brain in the mouse. *J Neuroimmunol* 47: 169–176

37   Banks WA, Kastin AJ (1993) Measurement of transport of cytokines across the blood-brain barrier. In: Conn PM, De Souza EB (eds): *Neurobiology of Cytokines*, Part A. Academic Press, San Diego, 67–77

38   Pan W, Banks WA, Fasold MB, Bluth J, Kastin AJ (1998) Transport of brain-derived neurotrophic factor across the blood-brain barrier. *Neuropharmacology* 37: 1552–1561

39   Banks WA, Kastin AJ (1989) Quantifying carrier-mediated transport of peptides from the brain to the blood. In: Conn PM (ed): *Methods in Enzymology*, Vol. 168. Academic Press, San Diego, 652–660

40   Banks WA, Fasold MB, Kastin AJ (1997) Measurement of efflux rates from brain to blood. In: Irvine GB, Williams CH (eds): *Methods in Molecular Biology, Neuropeptide Protocols*, Vol. 73. Humana Press, Totowa, NJ, 353–360

41   Banks WA, Kastin AJ (1992) Bidirectional passage of peptides across the blood-brain barrier. *Prog Brain Res* 91: 139–148

42   Schinkel AH (1999) P-glycoprotein, a gatekeeper in the blood-brain barrier. *Adv Drug Deliv Rev* 36: 179–194

43   Terasaki T, Hosoya K (1999) The blood-brain barrier efflux transporters as a detoxifying system for the brain. *Adv Drug Deliv Rev* 36: 195–209

44   Tsuji A, Tamai I (1999) Carrier-mediated or specialized transport of drugs across the blood-brain barrier. *Adv Drug Deliv Rev* 36: 277–290

45   King M, Su W, Chang A, Zukerman A, Pasternak GW (2001) Transport of opioids from the brain to the periphery by P-glycoprotein: peripheral actions of central drugs. *Nature Neuroscience* 4: 268–274

46   Kastin AJ, Fasold MB, Zadina JE (2002) Endomorphins, Met-enkephalin, Tyr-MIF-1, and the P-glycoprotein efflux system. *Drug Metab Disp* 30: 231–234

47   Deguchi Y, Naito T, Yuge T, Furukawa H, Yamada S, Pardidge WM, Kimura R (2000) Blood-brain barrier transport of ^{125}I-labeled basic fibroblast growth factor. *Pharm Res* 17: 63–69

48   Banks WA, Kastin AJ, Akerstrom V (1997) HIV-1 protein GP120 crosses the blood-brain barrier: role of adsorptive endocytosis. *Life Sci* 61: PL119–PL125

49   Banks WA, Akerstrom V, Kastin AJ (1998) Adsorptive endocytosis mediates the passage of HIV-1 across the BBB: evidence for a post-internalization coreceptor. *J Cell Sci* 111: 533–540

50   Maresh GA, Kastin AJ, Brown TT, Zadina JE, Banks WA (1999) Peptide transport system 1 (PTS-1) for Tyr-MIF-1 and Met-enkephalin differs from the receptors for either. *Brain Res* 839: 336–340

51   Barr VA, Lane K, Taylor SI (1999) Subcellular localization and internalization of the four human leptin receptor isoforms. *J Biol Chem* 274: 21416–21424

52   Golden PL, Maccagnan TJ, Pardridge WM (1997) Human blood-brain barrier leptin receptor. *J Clin Invest* 99: 14–18

53   Baura GD, Foster DM, Porte Jr, D, Kahn SE, Bergman RN, Cobelli C, Schwartz MW (1993) Saturable transport of insulin from plasma into the central nervous system of dogs *in vivo*: a mechanism for regulated insulin delivery to the brain. *J Clin Invest* 92: 1824–1830

54   Pan W, Kastin AJ (2002) TNFα transport across the blood-brain barrier is abolished in receptor knockout mice. *Exp Neurol* 174: 193–200

55   Kastin AJ, Pan W, Maness LM, Koletsky RJ, Ernsberger P (1999) Decreased transport of leptin across the blood-brain barrier in rats lacking the short form of the leptin receptor. *Peptides* 20: 1449–1453

56   Felgenhauer K (1974) Protein size and cerebrospinal fluid composition. *Klin Wochenschr* 52: 1158–1164

57   Fenstermacher JD (1981) Methods for quantifying the transport of drugs across blood-brain barrier systems. *Pharmacol Ther* 14: 217–248

58   Kastin AJ, Akerstrom V, Hackler L, Pan W (2001) Adrenomedullin and the blood-brain-barrier. *Hormones Metab Res* 33: 19–25

59   Kastin AJ, Akerstrom V, Pan W (2002) Interactions of glucagon-like peptide-1 (GLP-1) with the blood-brain barrier. *J Molec Neurosci* 18: 7–14

60   Kastin AJ, Castellanos PF, Fischman AJ, Proffitt JK, Graf MV (1984) Evidence for peptide aggregation. *Pharmacol Biochem Behav* 21: 969–973

61   Kastin AJ, Akerstrom V, Hackler L (2000) Agouti-related protein (83-132) aggregates and crosses the blood-brain barrier slowly. *Metabolism* 49: 1444–1448

62   Kastin AJ, Akerstrom V, Hackler L, Zadina JE (2000) Phe[13],Tyr[19]-Melanin-concentrating hormone and the blood-brain barrier: role of protein binding. *J Neurochem* 74: 385–391

63   Kastin AJ, Akerstrom V, Hackler L (2001) Food deprivation decreases blood galanin-like peptide and its rapid entry into the brain. *Neuroendocrinology* 74: 423–432

64   Banks WA, Kastin AJ (1995) Regional variation in transport of pancreatic polypeptide across the blood-brain barrier. *Pharmacol Biochem Behav* 51: 139–147

65   Banks WA, Kastin AJ, Maness LM, Huang W, Jaspan JB (1995) Permeability of the blood-brain barrier to amylin. *Life Sci* 57: 1993–2001

65a   Kastin AJ, Akerstrom V (1999) Entry of CART into brain is rapid but not inhibited by excess CART or leptin. *Am J Physiol* 277: E901–E904

66   Martins JM, Kastin AJ, Banks WA (1996) Unidirectional specific and modulated brain to blood transport of corticotropin-releasing hormone. *Neuroendocrinology* 63: 338–348

67   Martins JM, Banks WA, Kastin AJ (1997) Acute modulation of the active carrier-mediated brain-to-blood transport of corticotropin-releasing hormone. *Am J Physiol* 272: E312–E319

68   Martins JM, Banks WA, Kastin AJ (1997) Transport of CRH from mouse brain directly affects peripheral production of β-endorphin by the spleen. *Am J Physiol* 273: E1083–E1089

69   Kastin AJ, Akerstrom V (2002) Differential interactions of urocortin/corticotropin-releasing hormone peptides with the blood-brain barrier. *Neuroendocrinology* 75: 367–374

70   Kyrkouli SE, Stanley BG, Seirafi RD, Leibowitz SF (1990) Stimulation of feeding by galanin: anatomical localization and behavioral specificity of this peptide's effects in the brain. *Peptides* 11: 995–1001

71   Banks WA, Jaspan JB, Kastin AJ (1997) Selective, physiological transport of insulin across the blood-brain barrier: novel demonstration by species-specific radioimmunoassays. *Peptides* 18: 1257–1262

72   Banks WA, Jaspan JB, Huang W, Kastin AJ (1997) Transport of insulin across the blood-brain barrier: saturability at euglycemic doses of insulin. *Peptides* 18: 1423–1429

73   Banks WA, Jaspan JB, Kastin AJ (1997) Effect of diabetes mellitus on the permeability of the blood-brain barrier to insulin. *Peptides* 18: 1577–1584

74   Zhang Y, Proenca R, Maffei M, Barone M, Leopold L, Friedman JM (1994) Positional cloning of the mouse obese gene and its human analogue. *Nature* 372: 425–432

75   Wu-Peng XS, Chua SC Jr, Okada N, Liu S-M, Nicolson M, Leibel RL (1997) Phenotype of obese Koletsky (f) rat due to Tyr763 stop mutation in the extracellular domain of the leptin receptor (lepr). *Diabetes* 46: 513–518

76   Van Heek M, Compton DS, France CF, Tedesco RP, Fawzi AB, Graziano MP, Sybertz EJ, Strader CD, Davis HR Jr (1997) Diet-induced obese mice develop peripheral, but not central, resistance to leptin. *J Clin Invest* 99: 385–390

77   Schwartz MW, Peskind E, Raskind M, Boyko EJ, Porte D, Jr. (1996) Cerebrospinal fluid leptin levels: relationship to plasma levels and to adiposity in humans. *Nature Med* 2: 589–593

78   Caro JF, Kolaczynski JW, Nyce MR, Ohannesian JP, Opentanova I, Goldman WH, Lynn RB, Zhang P-L, Sinha MK, Considine RV (1996) Decreased cerebrospinal-fluid/serum leptin ratio in obesity: a possible mechanism for leptin resistance. *Lancet* 348: 159–161

79   Koistinen HA, Karonen S-L, Iivanainen M, Koivisto VA (1998) Circulating leptin has saturable transport into intrathecal space in humans. *Eur J Clin Invest* 28: 894–897

80   Banks WA, Clever CM, Farrell CL (2000) Partial saturation and regional variation in the blood to brain transport of leptin in normal weight mice. *Am J Physiol* 278: E1158–E1165

81   Maness LM, Banks WA, Kastin AJ (2000) Persistence of blood-to-brain transport of leptin in obese leptin-deficient and leptin receptor-deficient mice. *Brain Res* 873: 165–167

82   Kastin AJ, Akerstrom V (2000) Mahogany (1377-1428) enters brain by a saturable transport system. *J Pharmacol Exp Ther* 294: 633–636

83   Dinulescu DM, Fan W, Boston BA, McCall K, Lamoreux ML, Moore KJ, Montagno J, Cone RD (1998) Mahogany (mg) stimulates feeding and increases basal metabolic rate independent of its suppression of agouti. *Proc Natl Acad Sci USA* 95: 12707–12712

84   Wilson JF, Anderson S, Snook G, LLewellyn KD (1984) Quantification of the permeability of the blood-CSF barrier to α-MSH in the rat. *Peptides* 5: 681–685

85   Banks WA, Kastin AJ (1995) Permeability of the blood-brain barrier to melanocortins. *Peptides* 16: 1157–1161

86   Forbes S, Bui S, Robinson BR, Hochgeschwender U, Brennan MB (2001) Integrated control of appetite and fat metabolism by the leptin-proopiomelanocortin pathway. *Proc Natl Acad Sci USA* 98: 4233–4237

87   Kastin AJ, Akerstrom V (1999) Nonsaturable entry of neuropeptide Y into the brain. *Am J Physiol* 276: E479–E482

88   Banks WA, Kastin AJ (1985) Peptides and the blood-brain barrier: lipophilicity as a predictor of permeability. *Brain Res Bull* 15: 287–292

89   Kastin AJ, Akerstrom V (1999) Orexin A but not orexin B rapidly enters brain from blood by simple diffusion. *J Pharmacol Exp Ther* 289: 219–223

90   Kastin AJ, Akerstrom V, Pan W (2000) Activation of urocortin transport into brain by leptin. *Peptides* 21: 1811–1818

91   Kastin AJ, Akerstrom V (2001) Pretreatment with glucose increases entry of urocortin into mouse brain. *Peptides* 22: 829–834

92   Kastin AJ, Nissen C, Nikolics K, Medzihradszky K, Coy DH, Teplan I, Schally AV (1976) Distribution of [^{3}H]α-MSH in rat brain. *Brain Res Bull* 1: 19–26
93   Kastin AJ, Nissen C, Schally AV, Coy DH (1976) Blood-brain barrier, half-time disappearance, and brain distribution for labeled enkephalin and a potent analog. *Brain Res Bull* 1: 583–589

Progress in Drug Research, Vol. 61
(L. Prokai and K. Prokai-Tatrai, Eds.)
© 2003 Birkhäuser Verlag, Basel (Switzerland)

Intracranial delivery of proteins and peptides as a therapy for neurodegenerative diseases

By Richard Grondin,
Zhiming Zhang, Yi Ai,
Don M. Gash and
Greg A. Gerhardt

Department of Anatomy &
Neurobiology and the
Morris K. Udall Parkinson's Disease
Research Center of Excellence
305 Davis-Mills Bldg
University of Kentucky Medical
Center
Lexington, KY 40536-0098
USA
<rcgron0@mky.edu>

Richard C. Grondin

was born in 1969 in Theford-Mines, Quebec, Canada. He graduated from Laval University in Quebec City, Canada in December 1997 with a Ph.D. in neurobiology. The focus of his Ph.D. thesis was to better understand the mechanisms underlying levodopa-induced dyskinesia, using the MPTP-exposed nonhuman primate model of Parkinson's disease. In January of 1998, he moved to Lexington, Kentucky to join the laboratory of Dr. Don M Gash. He was awarded a postdoctoral fellowship from the Fond de Recherche en Santé du Québec (FRSQ) to complete a three-year training. He currently holds the position of Research Assistant Professor in the Departement of Anatomy & Neurobiology at the University of Kentucky Medical Center. His research focuses primarly on studying the regenerative potential of glial cell line-derived neurotrophic factor (GDNF) in nonhuman primate models of aging and Parkinson's disease. Dr. Grondin is the author/co-author of over 40 peer-reviewed or invited papers.

Zhiming Zhang

was born in Beijing, China in 1955. He graduated from the Capital University of Medical Science in Beijing, China in 1983. He finished his residency of neurosurgery in Xuan Wu Hospital in 1989 and was promoted to attending doctor of neurosurgery before joining Dr. Don M. Gash's group in July of 1989. He currently holds the position of Research Associate Professor at the University of Kentucky. In the past year, he has developed techniques to measure pharmacologically evoked blood oxygen level-dependent changes in the basal ganglia of MRI-adapted, awake rhesus monkeys using functional MRI. Currently, he is quantifying the changes inside the brain accompanying chronic GDNF infusion by utilizing the technique he has developed. He has published, as author or co-author, over 30 peer-reviewed or invited papers.

Yi Ai

was trained as a neurologist at the Capital University of Medical Science in Beijing, China. After earning her M.D. in December of 1983, she finished her residency of neurology in Xuan Wu Hospital and was promoted to Chief Resident prior to joining Dr. Don M. Gash's group at the University of Rochester in 1990. Dr. Ai currently works at the University of Kentucky as a scientist. She has been involved in neurodegenerative disease research such as Parkinson's and Alzheimer's disease since she graduated from medical school. Her current research focuses on cellular and molecular changes occurring in nonhuman primate models of aging and Parkinson's disease following treatment with glial cell line-derived neurotrophic factor (GDNF).

Don M. Gash

received his Ph.D in Biological Sciences from Dartmouth College in 1975. Following two years of postdoctoral training at the University of Southern California, College of Medicine, he joined the faculty of the Department of Neurobiology and Anatomy at the University of Rochester, Rochester, New York. In 1993, he was appointed Professor and Chair of Anatomy and Neurobiology in the College of Medicine at the University of Kentucky. He currently holds the Alumni Endowed Chair in the department. Current administrative duties include Department Chair (Anatomy and Neurobiology) and Director of the Magnetic Resonance Imaging and Spectroscopy Center. He has published over 180 papers, invited reviews and book chapters.

Greg A. Gerhardt

received his Ph.D. in chemistry with Dr. Ralph N. Adams at the University of Kansas. He carried out postdoctoral training in pharmacology, psychiatry and neuroscience at the University of Colorado Health Sciences Center. He is currently Professor (with tenure), Departments of Anatomy & Neurobiology, Neurology and Psychiatry, and Director of the Morris K. Udall Parkinson's Disease Research Center of Excellence at the University of Kentucky Chandler Medical Center, Lexington, Kentucky. He is also the Director of the Center for Sensor Technology (CenSeT) and the American Editor-in-Chief of the Journal of Neuroscience Methods. He is funded through grant support from the National Institute on Aging, National Institute on Mental Health, National Institute on Neurological Disorders and Stroke, DARPA, DOD and the National Science Foundation. He has received numerous awards including a recent Level II Research Scientist Development Award from NIMH (2000–2005) and he received an Outstanding Alumni Award from Truman State University in 1999. He has published over 200 original papers and book chapters and over 250 scientific abstracts. His research focuses on Parkinson's disease, normal aging and the repair of damaged dopamine neurons in the basal ganglia of the brain using growth factors such as GDNF. In addition, his laboratory develops technologies to directly measure chemical communication in the living brain.

Contents

Key words

Parkinson's disease, aging, glial cell line-derived neurotrophic factor, rhesus monkey, substantia nigra, putamen, lateral ventricle, programmable pumps.

Glossary of abbreviations

GDNF, glial cell line-derived neurotrophic factor; GPI, glycosyl-phosphatidylinositol; HVA, homovanillic acid; MPTP, 1-methyl-4-phenyl-1,2,3,6-tetrahydropiridine; MRI, magnetic resonsance imaging; O.D., outer diameter; PD, Parkinson's disease; TGF, transforming growth factor.

1　Introduction

Parkinson's disease (PD) is characterized by a progressive degeneration of the substantia nigra pars compacta dopamine neurons that innervate the striatum [1]. The loss of striatal dopamine and the consequent dysfunction of the nigrostriatal pathway lead to the cardinal symptoms of PD: resting tremor, cogwheel rigidity, bradykinesia (or difficulty to initiate movement) and loss of postural reflex. Current pharmacological treatment strategies for PD aim at replacing striatal dopamine using the dopamine precursor levodopa or dopamine receptor agonists, or both. Such treatments provide symptomatic relief, but do not slow or halt continued degeneration of nigral dopaminergic neurons. Therefore, interventions that could potentially slow or reverse the progression of neuronal degeneration would benefit PD patients. One such approach involves trophic factor administration. Trophic factors are proteins with enormous therapeutic potential in the treatment of neurodegenerative diseases. However, delivery of these proteins to the brain remains a challenge. Theoretically, a variety of different routes of administration could be used to deliver neurotrophic factors to the diseased brain, including intracranial (e.g. intraparenchymal), intranasal, intrathecal and parenteral (e.g. intravenous, subcutaneous or intramuscular) administration. Parenteral administration requires the drug to enter the brain from the systemic circulation by crossing the blood-brain barrier. Because trophic factors have poor blood-brain barrier permeability, access to sites of action within the brain is minimal *via* parenteral administration (i.e. < 1% of the injected dose) [2]. Along these lines, the use of intrathecal administration as a treatment for neurodegenerative diseases of the brain is not optimal, as therapeutic concentrations of the drug delivered at the cisternal or lumbar spinal cord level is rarely reached in other cerebrospinal fluid spaces [2]. Intranasal administration offers a method for bypassing the blood-brain barrier and contributes to better central nervous system penetration of neurotrophic factors [2]. However, the nasal route does not allow for focal delivery of these potent molecules and high doses of trophic factors may be necessary to elicit therapeutic effects. On the other hand, direct intraparenchymal administration of trophic factors into the brain (e.g. nigrostriatal pathway) offers a high degree of targeting, while reducing the risks of unwanted side-effects by limiting systemic exposure. Different methods allowing for direct intraparenchymal delivery have been studied in animals models, such as implantation of encapsulated cells genetically engineered to produce and release trophic factors locally [3]. Additional methods are currently being investigated in non-human primates, including viral vector-mediated delivery and the use of computer-controlled infusion pumps. In this chapter, we will review promising data recently obtained in non-human primates modeling PD that could help lay the foundation for prolonged, intraparenchymal delivery of trophic factors or other large molecules into the human brain.

2   Neurotrophic factors in neurodegenerative diseases

Neurotrophic factors are naturally occurring proteins required for neuronal differentiation, guidance and survival during development, as well as for the maintenance of the adult nervous system. Classically, a neurotrophic factor is produced and secreted by target cells, be they nerve cells or other cells, and then taken up by the innervating nerve terminals to exert both local effects and, *via* retrograde axonal transport, trophic effects in the nerve cell bodies [4]. Considerable effort has been devoted to the search for neurotrophic factors with survival-promoting activities on midbrain dopaminergic neurons that could potentially be of therapeutic value in the treatment of PD.

There are currently more than 20 trophic factors that have been identified, showing potential for use in a variety of neurodegenerative diseases including PD [2, 5]. Although there is little evidence that deficiencies of trophic factors are associated with the etiology of PD [6], several factors have been shown to produce significant beneficial effects on dopamine neurons in culture and in animal models [5]. Those producing effects on dopamine neurons *in vitro* include, but are not limited to, fibroblast growth factor, epidermal growth factor, transforming growth factor-α (TGF-α), ciliary neurotrophic factor, platelet-derived growth factor, brain-derived neurotrophic factor, and glial cell line-derived neurotrophic factor (GDNF) [2, 5]. However, of all the factors investigated to date, GDNF is the only trophic factor, which has been shown to dramatically protect and enhance the function of dopamine neurons in animal models [7, 8].

3   GDNF properties

GDNF was isolated and purified from the conditioned medium of cultured rat glial cells from the B49 cell line [9]. GDNF is a heparin binding protein and acts as a disulfide-bonded dimer. Each portion of the mature protein consists of 134 amino acid residues, with 93% identity between the human and rat sequences. The naturally occurring dimer has a molecular weight of ~30 kDa [10]. The subsequent purification and cloning of new trophic factors related to GDNF, termed neurturin [11], persephin [12] and artemin [13], has established the existence of a new family of neurotrophic factors structurally similar to the transforming growth factor-β (TGF-β) superfamily.

GDNF is widely expressed throughout the body in many neuronal (e.g. striatum, cerebellum, cortex) and non-neuronal tissues (e.g. kidney, gut) [14–18]. GDNF uses a multisubunit receptor system consisting of a glycosyl-phosphatidylinositol (GPI)-anchored membrane protein, termed GFRα-1, ("α" subunit), that can bind GDNF and facilitate its interaction with the tyrosine kinase Ret receptor ("β" subunit) [19, 20]. Three other GPI-anchored membrane proteins are now known, namely GFRα-2, GFRα-3 and GFRα-4 [21-23]. Mechanistically, little is known at the present time about the intracellular phosphorylation signaling mechanisms that occur *in vivo* following activation of Ret *via* GDNF or the related factors neurturin, persephin and artemin. Understanding the mechanism of action of GDNF is of great importance for its future use in treating neurodegenerative diseases of the brain such as PD.

4   GDNF effects in non-human primate models of Parkinson's disease

Data collected in cell culture [9, 10, 24] and in rodent models of PD [25–29] have shown that GDNF can be both neuroprotective and neurorestorative for the dopaminergic system, providing strong support for a role of GDNF in treating PD. Although crucial and informative, studies involving GDNF treatment in rodent models are limited in their relevance to the human. Rodents have a much smaller nervous system, which differs significantly in numerous neuroanatomical and neurochemical parameters from the human. In contrast, nonhuman primates possess a central nervous system and behavioral repertoire much closer to the human than the rodent.

4.1   Single or repeated intracerebral injections

Our group has carried out an extensive series of experiments to study the restorative effects of GDNF in nonhuman primates expressing hemiparkinsonian features as a result of infusions of 0.4 mg/kg 1-methyl-4-phenyl-1,2,3,6-tetrahydropiridine (MPTP) into the right carotid artery [30]. All the procedures performed on nonhuman primates in our laboratory are in strict accordance with the *NIH Guide for the Care and Use of Laboratory Animals* and are approved by institutional animal care and use committees. The surgeries are conducted under sterile field conditions in a sterile surgical suite accredited by the Association for Assessment and Accreditation of Laboratory Animal Care International (AAALACI).

Because GDNF does not cross the blood-brain barrier, a challenge to its clinical use is the difficulty associated with its delivery to the central nervous system. In our initial studies, sterile magnetic resonance imaging (MRI)-guided stereotaxic procedures were used to surgically deliver a single injection of GDNF directly into the right hemisphere of MPTP-lesioned hemiparkinsonian rhesus monkeys (*Macaca mulatta*) by one of three routes: intranigral (150 µg), intracaudate (450 µg) and intracerebroventricular (450 µg) [31]. Treatment was not started until two months post MPTP administration. GDNF recipients showed significant functional improvements from all three routes of administration by two weeks post-treatment, which continued for the remainder of the 4-week test period. In these experiments, improvements were found in three of the cardinal features of PD: bradykinesia, rigidity and postural instability. Further testing was carried out on the ventricular-treated animals to assess the ability of the animals to respond to three repeated dosings (100 or 450 µg), four weeks apart. GDNF administered every four weeks maintained functional recovery. Post-mortem analyses showed that dopamine levels in the midbrain and globus pallidus were twice as high on the lesioned side of GDNF-treated animals, and that nigral dopamine neurons were 20% larger on average, with an increase in fiber density compared to vehicle controls [31]. Follow up experiments have provided additional details about the antiparkinsonian actions of GDNF. Dose-dependent improvements in motor functions were seen in MPTP-lesioned hemiparkinsonian rhesus monkeys receiving monthly ventricular injections of 100–1000 µg GDNF. With monthly intraventricular injections, the optimal dose to produce a significant improvement in behavioral parameters was determined to be 300 µg [32].

Another group has studied the effects of GDNF administered intracerebrally in MPTP-treated common marmosets [33]. In their study, marmosets received parenteral administration of MPTP, which produced bilateral degeneration of the nigrostriatal pathway. Intraventricular injections of GDNF (10, 100 or 500 µg) were administered 9 and 13 weeks post MPTP treatment. This produced a dose-related improvement in locomotor activity and motor disabilities, which was significant following the 100 and 500 µg doses of GDNF.

Altogether, our studies [31, 32], as well as those of Costa and colleagues [33], have shown that efficacious GDNF dose levels using ventricular injection, spaced one month apart, ranged from 100 to 1000 µg per month. However, monthly ventricular injections of high dose GDNF increase the risk of inducing unwanted side effects, such as weight loss, which may limit the clinical applications of this approach [31, 32]. On the other hand, sustained delivery of low doses of GDNF into specific brain sites may produce the same or enhanced functional improvements as monthly ventricular injections with fewer side effects, by limiting systemic exposure. Moreover, due to the progressive nature of PD, sustained delivery of trophic factors may be necessary for optimal, long-term neuronal effects. Consequently, novel methods for sustained delivery of GDNF into the nigrostriatal pathway are currently being studied in non-human primates, including viral vector-mediated delivery and the use of computer-controlled infusion pumps.

4.2 Chronic, viral vector-mediated delivery

Kordower and colleagues [34] have recently carried out a series of studies to determine the effects of chronic GDNF administered by a lentiviral delivery approach (lenti-GDNF). In the first experiment, they investigated the effects of lenti-GDNF in four aged monkeys modeling some of the behavioral and cellular changes seen in the early stages of PD. In order to maximize the chance for an effect, each aged monkey (~25 years old) received a total of 6 simultaneous stereotaxic injections of lenti-GDNF into the right nigrostriatal dopaminergic system; i.e. caudate nucleus (n = 2 injections), putamen (n = 3 injections) and substantia nigra (n = 1 injection). This approach resulted in increased tissue levels of dopamine up to 140% in the right striatum, relative to control animals (i.e. lenti-βGal-treated). Also, stereological counts revealed an 85% increase in the number of tyrosine hydroxylase-immunoreactive nigral neurons in the right hemisphere.

In the second experiment, lenti-GDNF was also injected simultaneously into the right caudate nucleus, putamen and substantia nigra of five young adult rhesus monkeys, one week post MPTP administration into the right carotid artery. Chronic GDNF treatment reversed motor deficits seen in a hand reach task. In addition, nigrostriatal degeneration was prevented in the lenti-GDNF treated monkeys that received MPTP. Extensive GDNF expression with retrograde and anterograde transport was observed in the animals, with GDNF gene expression lasting for 8 months.

Overall, these experiments demonstrate that lentiviral vector-mediated delivery of GDNF can potently reverse structural and functional effects of dopamine deficiency in non-human primate models of aging and PD. However, two animals died within a week following lentiviral vector delivery. Although the authors attributed these deaths to MPTP toxicity rather than viral vector administration, caution is warranted regarding the safety of this approach in treating neurodegenerative diseases such as PD [35]. Also, there are important issues related to the development of reliable control mechanisms that would allow external regulation of the GDNF transgene expression after viral vector delivery into the brain, so that GDNF production could be reduced or completely turned-off to prevent or reverse adverse effects should they occur. Along these lines, inducible lentiviral vector systems containing the entire tetracycline-regulated system have been tested *in vitro* and *in vivo* in rats [36].

4.3 Chronic, computer-controlled delivery using programmable pumps

In many published animal studies, it is difficult to distinguish between the results from protection (injury prevention) and restoration (recovery after an injury) because GDNF treatment is initiated in the hours to days following a lesion, while the injury sequelae are still unfolding. Another issue is the titer of biologically available GDNF necessary to produce beneficial effects. For instance, while significant beneficial effects can be quantified on host dopamine neurons and neuronal processes after viral vector GDNF transfection [34], the levels of biologically available GDNF producing these effects are unclear. Additionally, techniques for controlling dosing and timing of viral vector-mediated delivery are in the developmental stage [37, 38].

Therefore, to determine the titer of biologically available GDNF necessary to produce beneficial effects, a second series of experiments was undertaken in our laboratory to study the safety and efficacy of chronically infusing computer-controlled doses of GDNF into the primate brain using implantable, programmable pumps (SynchroMed™ Model 8616-10, Medtronic Inc., Minneapolis MN). The pump can dispense drugs in a variety of ways (e.g. continuous or timed infusion) according to instructions received by radio-frequency from the SynchroMed™ Model 8820 computer. The pump is a round titanium disk, about one inch thick and three inches in diameter. The model 8616-10 pump contains a collapsible 10-ml reservoir and a self-sealing silicone septum through which a needle is inserted to refill the pump reservoir. It also contains a bacterial retentive filter (0.22 micron) through which the drug passes as it leaves the reservoir. The implantable pump is connected to a catheter made of polyurethane, stereotactically implanted into the brain [39]. A removable stylet inserted in the catheter lumen provides additional stiffness and control during placement. We have used four different types of catheters (1 mm outer diameter (O.D.)) for each of the three targets studied in our experiments: the lateral ventricle, the putamen and the substantia nigra. The ventricular catheter has a hole in the tip with two adjacent side holes for drug delivery (model 8770AS). Three different catheters have been used for intraparenchymal delivery. To chronically deliver GDNF into the putamen, we have used a porous tip catheter (model 8770IP3) or a multiport catheter with a radiodense-closed tip (model 8770IP24A). The multiport catheter is composed of six laser holes that are placed radially over each 90 degrees of the catheter's circumference over a longitudinal distance of 3 mm, for a total of 24 laser holes (0.0015″ or 37.5 µm in diameter). The most proximal set of radial holes are positioned 0.5 mm from the radiodense catheter tip. For placement into the substantia nigra, we have used catheters having a single opening (0.010″ or 250 µm in diameter) at the distal tip (model 8770IP1A).

4.3.1 MPTP rhesus monkey model of advanced Parkinson's disease

To assess the restorative actions of GDNF under conditions where neuroprotection would have only a minor role, the late stages of human PD can be modeled using rhesus monkeys displaying stable parkinsonian features for at least two months post MPTP administration [40–42]. In humans and non-human primates, MPTP induces behavioral features with numerous similarities to those found in idiopathic PD. For instance, MPTP-treated non-human primates display the cardinal symptomology of PD: bradykinesia, rigidity, balance and gait abnormalities [40, 41]. Additionally, histological and neurochemical alterations in the brain induced by MPTP administration also resemble those found in PD. Indeed, MPTP infusion through the right carotid artery results in an approximate 75% loss of dopamine neurons expressing the phenotypic marker tyrosine hydroxylase in the right substantia nigra and a greater than 99% depletion of dopamine in the right putamen [31]. These reductions are comparable with advanced human PD where cell counts typically show a 60–70% loss of nigral dopamine neurons [43] and 99% dopamine depletion in the putamen [44].

4.3.1.1 Intraventricular or intraputamenal delivery

Typically, the catheter is surgically positioned into the brain after a minimum of two months following the MPTP administration, when the parkinsonian features expressed by the animals have stabilized. The catheter is then seated in the groove of an L-shaped nylon device anchored against the skull using two nylon screws, and tunneled to the pump that is subcutaneously implanted in the lateral abdominal region [39].

Using this approach, we demonstrated that chronic infusions of nominally 7.5 or 22.5 µg/day GDNF into the lateral ventricle or the putamen, using programmable pumps, promotes restoration of the nigrostriatal dopaminergic system and significantly improves motor functions in rhesus monkeys with neural deficits modeling the terminal stages of PD [45]. The functional improvements were associated with a pronounced up-regulation and regeneration of nigral dopamine neurons and their processes innervating the striatum. When compared to vehicle recipients, these functional improvements were associated with: 1) > 30% bilateral increase in nigral dopamine neuron cell size, 2) > 20% bilateral increase in the number of nigral cells expressing the dopamine marker tyrosine hydroxylase, 3) > 70% and > 50% bilateral increase, respectively, in dopamine metabolite levels in the striatum and the pallidum, 4) 233% and 155% increase in dopamine levels in the periventricular striatal region and in the globus pallidus, respectively, on the lesioned side, and 5) a five-fold increase in tyrosine hydroxylase positive fiber density in the periventricular striatal region, on the lesioned side [45]. All of these effects from chronic administration of GDNF are greater than those previously seen from single injections of GDNF in MPTP-lesioned rhesus monkeys [31].

4.3.1.2 Intranigral delivery

In previous studies, GDNF has been administered to various sites, including simultaneous delivery to the striatum and substantia nigra of aged and MPTP-lesioned monkeys [34]. However, prolonged trophic factor administration into just the substantia nigra has not been evaluated and the effects are not known. Therefore, we have recently evaluated intranigral delivery of GDNF by measuring and analyzing the effects of GDNF on motoric behavior and tissue levels of dopamine and related metabolites in MPTP-lesioned hemi-parkinsonian monkeys [46]. In this study, stereotaxic procedures guided by MRI were used to implant a single-port catheter into the right substantia nigra of 10 rhesus monkeys, two months following the induction of parkinsonian features *via* MPTP administration. Using a SynchroMed™ programmable pump, five animals received chronic infusions of GDNF (7.5 µg/ day for 4 weeks and 22.5 µg/day for 8 weeks), in parallel with five control animals receiving vehicle.

This study showed that chronic intranigral delivery of 7.5 or 22.5 µg/day of exogenous GDNF, *via* programmable pumps, significantly improved motor functions in MPTP-lesioned rhesus monkeys [46]. As measured using the automated video tracking system, EthoVision® (Noldus, Asheville, NC), GDNF increased motor speed and home-cage activity levels (i.e. total distance traveled) up to 45% and 40% respectively, at the highest dose tested. Tissue punches taken from 4-mm-thick coronal sections through the basal ganglia after 12 weeks of GDNF administration showed that the motor improvements were associated with unilateral increases in dopamine levels up to 222% in the striatum on the lesioned right side, and with a bilateral increase in the globus pallidus externa up to 199%. The motor recovery was also associated with an increase in the dopamine metabolite, homovanillic acid (HVA), up to 132% in the striatum on the lesioned right side, and with a bilateral increase in the globus pallidus externa up to 171% [46].

Taken together, these data provide support that the chronic, intraparenchymal delivery of GDNF promotes restoration of the nigrostriatal dopaminergic system and significantly improves motor functions in MPTP-treated rhesus monkeys modeling the advanced stages of PD. Clearly, the prolonged and controlled delivery of GDNF into the brain, using programmable pumps, could be used to intervene in long-term neurodegenerative disease processes like PD.

4.3.2 Aged rhesus monkey model of early Parkinson's disease

The progressive slowing of motor functions and the development of parkinsonian signs are common features seen with advancing age in humans [47, 48] and in monkeys [49, 50]. Changes in the functional dynamics of dopamine release and regulation in the basal ganglia have been posited to contribute to age-related slowing of motor functions [51]. If so, interventions that up-regulate dopamine release in the basal ganglia should improve the motor deficits characterizing normal aging. It has been hypothesized that GDNF mediates its effects, at least in part, by regulating the neuronal excitability of midbrain dopaminergic neurons [52]. This GDNF-induced potentiation of neuronal excitability would result in an increase in dopamine release, leading to an enhancement in motor functions [52]. Little is known about the effects of GDNF on motoric behavior and dopaminergic functions in aged monkeys. Thus, a third series of experiments was designed to determine whether chronic intracerebral infusion of exogenous GDNF using programmable pumps can improve motoric deficits in aged monkeys, and if so, whether this improvement is associated with an increase in nigrostriatal dopamine functions. The aged rhesus monkeys used in these studies ranged in age from 20 to 27 years old, roughly equivalent to 60 to 80 years old in human age [53].

4.3.2.1 Intraventricular delivery

Using the programmable pump infusion approach, we reported that chronic infusions of 7.5 µg GDNF/day for two months into the right lateral ventricle of aged rhesus monkeys initially increased upper limb movement speed up to 40%, on an automated hand-reach task [54]. These effects were maintained for at least two months after replacing GDNF with vehicle, and further increased up to 50% following reinstatement of GDNF treatment for one month. In addition, upper limb motor performance times of the aged GDNF-treated animals (n = 5) recorded at the end of the study were similar to those of five young adult monkeys (8–12 years old). No changes were observed in the aged vehicle controls (n = 5). As measured by *in vivo* microdialysis, stimulus-evoked release of dopamine was significantly increased up to 130% in the right caudate nucleus and putamen, and up to 116% in both the right and left substantia nigra of the aged GDNF recipients compared to aged vehicle controls. Also, basal extracellular levels of dopamine were bilaterally increased up to 163% in the substantia nigra of the aged GDNF-treated animals [54].

The amplitude of amphetamine-induced release of dopamine observed in the putamen and in the substantia nigra of the GDNF-treated aged monkeys was similar to that of young adult rhesus monkeys [51]. The observation that GDNF can significantly improve motor functions and augment the functional capacity of dopamine neurons in aged rhesus monkeys to a level similar to that of young adults is of importance, considering that a major hallmark of aging in human is slowing of motor movements [47, 48]. Our GDNF washout data also suggest that continuous stimulation of GDNF receptors may not be required for GDNF to effectively improve motor functions. This observation could have a significant impact on future use of GDNF as a treatment for neurodegenerative diseases, such as PD. Consequently, additional studies are needed to determine whether improved motoric behavior following chronic GDNF infusion can be sustained for more than two months in the absence of GDNF, and whether the response would be different in aged *versus* MPTP-lesioned rhesus monkeys modeling PD.

4.3.2.2 Intraputamenal delivery

The effects of chronic intraputamenal infusion of GDNF were also studied in aged rhesus monkeys [55]. In this study, four animals received chronic infusions of GDNF and four received vehicle infusions into the right putamen *via* programmable pumps for eight weeks. Weekly videotaping was performed to record general motor performance and a food retrieval task was used to quantify fine and coarse upper limb motor performance. The GDNF-treated animals showed significant improvements in their overall motor performance in the last three weeks of the study compared to controls. Also, fine motor time of the upper limbs improved significantly in the GDNF-treated animals. After eight weeks of drug administration, the animals were euthanized and tissue punches were taken from the basal ganglia for measures of dopamine and dopamine metabolite levels. In the right putamen, GDNF infusion produced a 217% increase in HVA levels. In addition, dopamine levels increased by 50% in the right caudate nucleus and there were 122% and 76% increases in 3,4-dihydroxyphenylacetic acid levels in the right and left caudate nucleus, respectively. HVA levels were also increased by 212% in the right caudate nucleus. Finally, changes were seen in the right globus pallidus, with 390% and 171% increases in dopamine and HVA levels, respectively [55].

Taken together, our studies conducted in aged rhesus monkeys suggest that the effects of GDNF on dopaminergic functions in the basal ganglia may be responsible for the improvements in motor functions, and support the hypothesis that functional changes in dopamine release may contribute to motoric dysfunctions characterizing senescence. Aging nonhuman primates may represent an early stage model of PD, supporting that early intervention by trophic factor administration in PD patients may result in enhanced repair of dopamine neurons.

5 GDNF effects in human parkinsonian subjects

5.1 Ventricular delivery

Based on the promising studies of the effects of GDNF in animal models of PD, an initial clinical trial testing GDNF by ventricular delivery using an indwelling reservoir was carried out in 50 parkinsonian patients for 8 months [56]. While the doses of GDNF (25–4000 µg/month) were in excess of those employed for nonhuman primate studies, little therapeutic efficacy was observed in these parkinsonian patients. In fact, one 65-year-old patient with a 23-year history of PD who came to autopsy showed no significant effects of GDNF on dopamine neurons [57]. The problem may have been with the site and method of delivery; i.e. monthly injections of the trophic factor into the lateral ventricle. Sufficient titers of GDNF may not have diffused through the ventricular wall and brain parenchyma to the targeted dopamine neurons in the substantia nigra and their afferent projections to the putamen. As seen in Figure 1, this hypothesis has been born out from recent studies in our laboratory, which have demonstrated the limited penetration of GDNF through

A) GDNF: Intraventricular

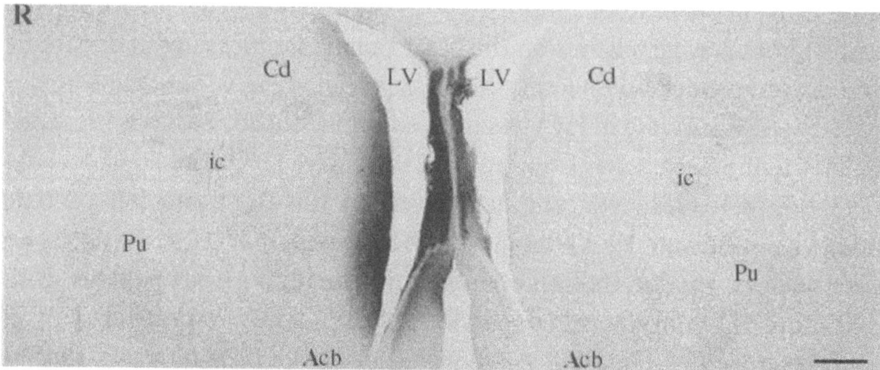

B) GDNF: Intraputamenal

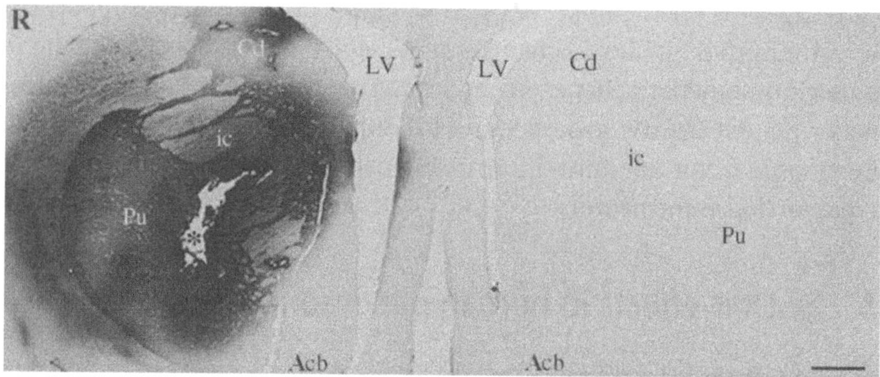

Figure 1.

As revealed in immunocytochemically stained sections, the spread of GDNF from the intraventricular (LV) catheter was limited in most cases to the right (R) paraventricular region of the caudate nucleus (Cd) and nucleus accumbens (Acb) ipsilateral to the catheter, diffusing into the parenchyma for distances up to 1.5–2 mm. Some bilateral diffusion was evident into the septum (panel A). In contrast, the spread of GDNF from the multiport intraputamenal catheters was more extensive, in roughly an elliptical pattern. The greatest radius of diffusion from the catheter ranged up to 11 mm in the rostral putamen (Pu) and caudate nucleus (panel B). Large regions of the rostral half of the putamen, internal capsule (ic) and caudate nucleus were filled with GDNF, with infused trophic factor extending as far caudally as the anterior globus pallidus (data not shown). Fiber tracts did not inhibit GDNF diffusion, as evident in GDNF staining in the internal capsule, corpus callosum and cortical white matter (panel B). No GDNF positive staining was evident in the vehicle recipients (data not shown). The scale bars indicate 2 mm in distance.

the ventricular wall as compared to the adequate diffusion of GDNF observed with intraparenchymal infusion into the striatum of nonhuman primates [58].

5.2 Chronic, computer-controlled delivery using programmable pumps

The excellent control achieved by pumps and infusion catheters have made this approach the best for initial trials in humans [59]. In a recent study conducted in England [60], five advanced PD patients with a previous history of good responses to levodopa underwent unilateral or bilateral insertion of drug infusion cannulae into the dorsal putamen (Fig. 2). Human recombinant GDNF was chronically infused *via* indwelling SynchroMed™ pumps implanted in the abdominal region (Fig. 2). Patients were assessed pre- and post-operatively according to the core assessment program for intracerebral transplantations (CAPIT), in order to document changes in disease severity and medication requirements [61]. The patients also underwent 18F-dopa positron emission tomography scans at baseline, 6- and 12-months after GDNF infusion to assess putamen dopamine terminal function and correlate this with any symptomatic benefits.

Chronic GDNF infusion resulted in improved motor function in all patients, reduction in "off"-time duration and severity, reduction in dyskinesias duration and severity, and a corresponding increase in good "on"-time duration. In four out of five patients reviewed at 12-months of drug administration, motor (Unified PD Rating Scale (UPDRS) III) and activities of daily living (UPDRS II) scores improved by 49% and 65%, respectively in the "off"-medication state, and by 33% and 43%, respectively in the "on"-medication state. "Off"-medication timed motor tests improved to pre-operative best "on"-medication times in the four cases, with a composite of the timed motor tests at 12-months showing improvements of 50% and 23% in the "off" and "on"-medication states, respectively. Chronic GDNF infusion was tolerated well in all patients and limited side effects were observed.

This is an exciting initial trial that is being followed up in the United States, at the University of Kentucky, with an FDA-approved Phase-I safety trial on the use of chronically administered GDNF in ten patients with

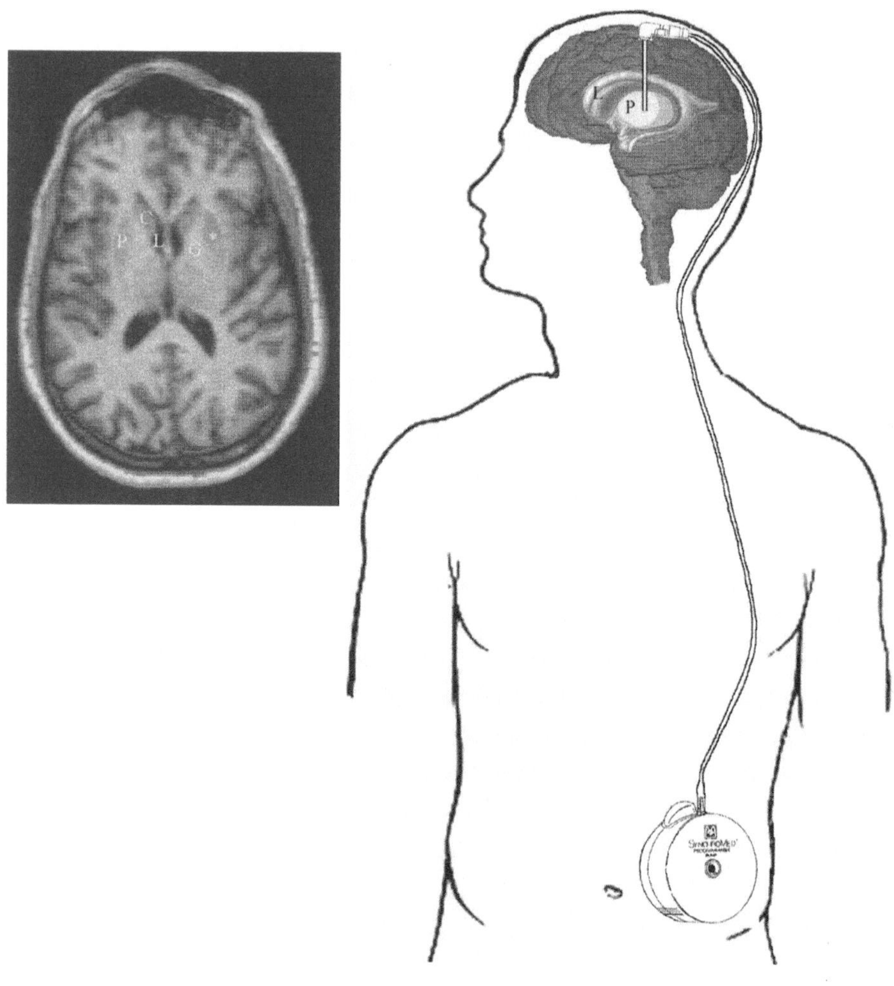

Figure 2.
The schematic illustrates the SynchroMed™ pump and catheter placement. The titanium-encased, biologically-compatible pump is surgically implanted in the subcutaneous layer of the abdomen and connected by tubing to a catheter stereotactically positioned into the putamen (*). The pump is refilled with GDNF by injections through the skin into a portal diaphragm. In addition, the rate and timing of delivery are noninvasively programmed from an external computer. C: caudate nucleus; G: globus pallidus; L: lateral ventricle; P: putamen.

advanced PD. These studies could not only help designing better treatment for GDNF, but could also lay the foundation for the chronic delivery of other molecules into the human brain. However, while the data from the recent intraparenchymal clinical trial in humans look encouraging [60], extensive blinded efficacy trials will need to be conducted before it can be determined if chronic treatment with GDNF or other trophic molecules will prove useful in treating patients with PD.

6   Summary

Parkinson's disease is characterized by a progressive degeneration of the substantia nigra pars compacta dopamine neurons that innervate the striatum. Unlike current treatments for PD, GDNF administration could potentially slow or halt the continued degeneration of nigral dopaminergic neurons. GDNF does not cross the blood-brain barrier and needs to be administered directly into the brain. Due to the progressive nature of PD, sustained delivery of trophic factors may be necessary for optimal, long-term neuronal effects. Novel methods for sustained delivery of GDNF into the nigrostriatal pathway are currently being studied in non-human primates, including computer-controlled infusion pumps. Using this approach, we have demonstrated that chronic infusions of nominally 7.5 or 22.5 µg/day GDNF into the lateral ventricle, the putamen or the substantia nigra, using programmable pumps, promotes restoration of the nigrostriatal dopaminergic system and significantly improves motor functions in MPTP-lesioned rhesus monkeys with neural deficits modeling the terminal stages of PD and in aged rhesus monkeys modeling the early stages of PD. Based on the promising studies of the chronic effects of GDNF in non-human primate models of PD, a study was recently conducted in England on five advanced PD patients. Chronic GDNF infusion into the dorsal putamen, *via* programmable pumps, resulted in improved motor function in all patients and limited side effects were observed. However, while the data from this intraparenchymal clinical trial in humans look encouraging, extensive blinded efficacy trials will need to be conducted before it can be determined if chronic treatment with GDNF or other trophic molecules will prove useful in treating patients with PD.

Acknowledgements

Supported by USPHS grants NS39787, AG13494, AG06434 and MH01245, and contracts with Amgen Inc., Thousand Oak, CA. We thank Medtronic

Inc., Minneapolis, MN, for providing the pumps and associated hardware and software used for chronic infusions.

References

1   Lang AE, Lozano AM (1998) Parkinson's disease. First of two parts. *N Engl J Med* 339: 1044–1053

2   Thorne RG, Frey WH II (2001) Delivery of neurotrophic factors to the central nervous system, pharmacokinetic considerations. *Clin Pharmacokinet* 40: 907–946

3   Zurn AD, Widmer HR, Aebischer P (2001) Sustained delivery of GDNF: towards a treatment for Parkinson's disease. *Brain Res Rev* 36: 222–229

4   Olson L (1996) Toward trophic treatment in parkinsonism: A primate step. *Nat Med* 2: 400–401

5   Collier TJ, Sortwell CE (1999) Therapeutic potential of nerve growth factors in Parkinson's disease. *Drugs Aging* 14: 261–87

6   Hornykiewicz O (1993) Parkinson's disease and the adaptive capacity of the nigrostriatal dopamine system: possible neurochemical mechanisms. *Adv Neurol* 60: 140–147

7   Gash DM, Zhang Z, Gerhardt GA (1998) Neuroprotective and neurorestorative properties of GDNF. *Ann Neurol* 44: S121–S125

8   Grondin R, Gash DM (1998) Glial cell line-derived neurotrophic factor (GDNF) as a drug candidate for the treatment of Parkinson's disease. *J Neurol* 245 (Suppl 3): S35–S42

9   Lin L-FH, Doherty DH, Lile JD, Bektesh S, Collins F (1993) GDNF: A glial cell line-derived neurotrophic factor for midbrain dopaminergic neurons. *Science* 260: 1130–1132

10 Lin L-FH, Zhang TJ, Collins F, Armes LG (1994) Purification and initial characterization of rat B49 glial cell line-derived neurotrophic factor. *J Neurochem* 63: 758–768

11 Kotzbauer PT, Lampe PA, Heuckeroth RO, Golden JP, Creedon DJ, Johnson EM Jr, Milbrandt J (1996) Neurturin, a relative of glial cell line-derived neurotrophic factor. *Nature* 384: 467–470

12 Milbrandt J, de Sauvage FJ, Fahrner TJ, Baloh RH, Leitner ML, Tansey MG, Lampe PA, Heuckeroth RO, Kotzbauer PT, Simburger KS et al (1998) Persephin, a novel neurotrophic factor related to GDNF and neurturin. *Neuron* 20: 245–253

13 Baloh RH, Tansey MG, Lampe PA, Fahrner TJ, Enomoto H, Simburger KS, Leitner ML, Toshiyuki A, Johnson EM Jr, Milbrant J (1998) Artemin, a novel member of the GDNF ligand family, supports peripheral and central neurons and signals through the GFRalpha3-Ret receptor complexe. *Neuron* 21: 1291–1302

14 Strömberg I, Bjorklund L, Johansson M, Tomac A, Collins F, Olson L, Hoffer B, Humpel C (1993) Glial cell line-derived neurotrophic factor is expressed in the developing but not adult striatum and stimulates developing dopamine neurons *in vivo*. *Exp Neurol* 124: 401–412

15 Springer JE, Mu X, Bergmann, LW, Trojanoski JQ (1994) Expression of GDNF mRNA in rat and human nervous tissue. *Exp Neurol* 127: 167–170

16 Suter-Crazzolara C, Unsicker K (1994) GDNF is expressed in two forms in many tissues outside the CNS. *NeuroReport* 5: 2486–2488

17 Choi-Lundberg DL, Bohn MC (1995) Ontogeny and distribution of glial cell line-derived neurotrophic factor (GDNF) mRNA in rat. *Dev Brain Res* 85: 80–88

18 Trupp M, Rydén M, Jörnavall C, Funakoshi H, Timmusk T, Arenas E, Ibanez CF (1995)

Peripheral expression and biological activities of GDNF, a new neurotrophic factor for avian and mammalian peripheral neurons. *J Cell Biol* 130: 137–148

19 Jing S, Wen D, Yu Y, Holst PL, Luo Y, Fang M, Tamir R, Antonio L, Hu Z, Cupples R et al (1996) GDNF-induced activation of the Ret protein tyrosine kinase is mediated by GDNFR-alpha, a novel receptor for GDNF. *Cell* 85: 1113–1124

20 Treanor JJS, Goodman L, de Sauvage F, Stone DM, Poulsen KT, Beck CD, Gray C, Armanini MP, Pollock RA, Hefti F et al (1996) Charaterization of a multicomponent receptor for GDNF. *Nature* 382: 80–83

21 Baloh RH, Tansey MG, Golden JP, Creedon DJ, Heuckeroth RO, Keck CL, Zimonjic DB, Popescu NC, Johnson EM Jr, Milbrandt J (1997) TrnR2, a novel receptor that mediates neurturin and GDNF signalling through Ret. *Neuron* 18: 793–802

22 Naveilhan P, Baudet C, Mikaels A, Shen L, Westphal H, Ernfors P (1998) Expression and regulation of GFRalpha3, a glial cell line-derived neurotrophic factor family receptor. *Proc Natl Acad Sci USA* 95: 1295–1300

23 Thompson J, Doxakis E, Pinon LGP, Strachan P, Buj-Bello A, Wyatt S, Buchman VL, Davies AM (1998) GFRalpha4, a new GDNF family receptor. *Mol Cell Neurosci* 11: 117–126

24 Hou J-GG, Lin L-FH, Mytilinenou C (1996) Glial cell line-derived neurotrophic factor exerts neurotrophic effects on dopaminergic neurons *in vitro* and promotes their survival and regrowth after damage by 1-methyl-4-phenylpyridinium. *J Neurochem* 66: 74–82

25 Hoffer BJ, Hoffman A, Bowenkamp K, Huettl P, Hudson J, Martin D, Lin L-FH, Gerhardt G (1994) Glial cell line-derived neurotrophic factor reverses toxin-induced injury to midbrain dopaminergic neurons *in vivo*. *Neurosci Lett* 182: 107–111

26 Bowenkamp KE, Hoffman AF, Gerhardt GA, Henry MA, Biddle PT, Hoffer BJ, Granholm A-CE (1995) Glial cell line-derived neurotrophic factor supports survival of injured midbrain dopaminergic neurons. *J Comp Neurol* 355: 479–489

27 Tomac A, Lindqvist E, Lin L-FH, Ogren SO, Young D, Hoffer BJ, Olson L (1995) Protection and repair of the nigrostriatal dopaminergic system by GDNF *in vivo*. *Nature* 373: 335–339

28 Kearns C, Gash DM (1995) GDNF protects nigral dopamine neurons against 6-hydroxydopamine *in vivo*. *Brain Res* 672: 104–111

29 Lapchak PA, Miller PJ, Collins F, Jiao S (1997) Glial cell line-derived neurotrophic factor attenuates behavioural deficits and regulates nigrostriatal dopaminergic and peptidergic markers in 6-hydroxydopamine-lesioned adult rats: Comparison of intraventricular and intranigral delivery. *Neuroscience* 78: 61–72

30 Ovadia A, Zhang Z, Gash DM (1995) Increased susceptibility to MPTP toxicity in middle-aged rhesus monkeys. *Neurobiol Aging* 16: 931–937

31 Gash DM, Zhang Z, Ovadia A, Cass WA, Yi A, Simmerman L, Russell D, Martin D, Lapchak PA, Collins F et al (1996) Functional recovery in parkinsonian monkeys treated with GDNF. *Nature* 380: 252–255

32 Zhang Z, Miyoshi Y, Lapchak PA, Collins F, Hilt D, Lebel C, Kryscio R, Gash DM (1997) Dose response to intraventricular glial cell line-derived neurotrophic factor administration in parkinsonian monkeys. *J Pharmacol Exp Ther* 282: 1396–1401

33 Costa S, Iravani MM, Pearce RK, Jenner P (2001) Glial cell line-derived neurotrophic factor concentration dependently improves disability and motor activity in MPTP-treated common marmosets. *Eur J Pharmacol* 412: 45–50

34 Kordower JH, Emborg ME, Bloch J, Ma SY, Chu Y, Leventhal L, McBride J, Chen E-Y, Palfi

S, Zion Roitberg B et al (2000) Neurodegeneration prevented by lentiviral vector delivery of GDNF in primate models of Parkinson's disease. *Science* 290: 767–773

35   Bjorklund A, Kirik D, Rosenblad C, Georgievska B, Lundberg C, Mandel RJ (2000) Towards a neuroprotective gene therapy for Parkinson's disease: use of adenovirus, AAV and lentivirus vectors for gene transfer of GDNF to the nigrostriatal system in the rat Parkinson model. *Brain Res* 886: 82–98

36   Kafri T, van Praag H, Gage FH, Verma IM (2000) Lentiviral vectors: regulated gene expression. *Mol Ther* 1: 516–521

37   Olson L (2000) Combating Parkinson's disease–step three. *Science* 290: 721–724

38   Bjorklund A, Lindvall A (2000) Parkinson's disease gene therapy moves towards the clinic. *Nature Med* 6: 1207–1208

39   Grondin R, Zhang Z, Elsberry D, Gerhardt GA, Gash DM (2001) Chronic intracerebral delivery of trophic factors *via* a programmable pump as a treatment for parkinsonism. In: MM Mouradian (ed): *Parkinson's disease-methods & protocols*. Humana Press, Totowa, NJ, 62: 257–267

40   Bankiewicz KS, Oldfield EH, Chiueh CC, Doppman JL, Jacobowitz DM, Kopin IJ (1986) Hemiparkinsonism in monkeys after unilateral internal carotid artery infusion of 1-methyl-4-phenyl-1,2,3,6-tetrahydropyridine. *Life Sci* 39: 7–16

41   Smith RD, Zhang Z, Kurlan R, McDermott M, Gash DM (1993) Developing a stable bilateral model of parkinsonism in rhesus monkeys. *Neuroscience* 52: 7–16

42   Emborg-Knott ME, Domino EF (1998) MPTP-induced hemiparkinsonism in nonhuman primates 6–8 years after a single unilateral intracarotid dose. *Exp Neurol* 152: 214–220

43   Jellinger K (1986) Overview of morphological changes in Parkinson's disease. *Adv Neurol* 45: 1–17

44   Kish SJ, Shannak K, Hornykiewicz O (1988) Uneven pattern of dopamine loss in the striatum of patients with idiopathic Parkinson's disease. Pathophysiology and clinical implications. *N Eng J Med* 318: 876–880

45   Grondin R, Zhang Z, Cass WA, Ai Y, Maswood N, Andersen AH, Elsberry DD, Klein MC, Gerhardt GA, Gash DM (2002) Chronic, controlled GDNF infusion promotes structural and functional recovery in advanced parkinsonian monkeys. *Brain* 125: 2191–2201

46   Grondin R Zhang Z, Ai Y, Surgener SP, Loveland AD, Gerhardt GA, Gash DM (2002) Chronic, pulsatile delivery of GDNF into the substantia nigra increases motor speed and dopamine levels in the basal ganglia of MPTP-treated parkinsonian monkeys. *Soc Neurosci Abst* 28: 691.3

47   Bennett DA, Beckett LA, Murray AM, Shannon KM, Goetz CG, Pilgrim DM, Evans DA (1996) Prevalence of parkinsonian signs and associated mortality in a community population of older people. *New Engl J Med* 334: 71–76

48   Smith C, Umberger G, Manning EL, Slevin JT, Wekstein DR, Markesbery WR, Zhang Z, Gerhardt GA, Gash DM (1999) A critical decline in fine motor hand movements in human aging. *Neurology* 53: 1458–1461

49   Emborg ME, Ma SY, Mufson EJ, Levey AI, Taylor MD, Brown WD, Holden JE, Kordower JH (1998) Age-related declines in nigral neuronal function correlate with motor impairments in rhesus monkeys. *J Comp Neurol* 401: 253–265

50   Zhang Z, Andersen A, Smith C, Grondin R, Gerhardt GA, Gash DM (2000) Motor slowing and parkinsonian signs in aging rhesus monkeys mirror human aging. *J Gerontol Biol Sci* 55A: 473–480

51   Gerhardt GA, Cass WA, Yi A, Zhang Z, Gash DM (2002) Changes in somatodendritic but not terminal dopamine regulation in aged rhesus monkeys. *J Neurochem* 80: 168–177

52   Yang F, Feng L, Zheng F, Johnson SW, Du J, Shen L, Wu C, Lu B (2001) GDNF acutely modulates excitability and A-type K^+ channels in midbrain dopaminergic neurons. *Nature Neurosci* 4: 1071–1078

53   Andersen AH, Zhang Z, Zhang M, Gash DM, Avison MJ (1999) Age-associated changes in rhesus CNS composition identified by MRI. *Brain Res* 829: 90–98

54   Grondin R, Cass WA, Zhang Z, Stanford JA, Gash DM, Gerhardt GA (2003) Glial cell line-derived neurotrophic factor increases stimulus-evoked dopamine release and motor speed in aged rhesus monkeys. *J Neurosci* 23: 1974–1980.

55   Maswood N, Grondin R, Zhang Z, Stanford JA, Surgener SP, Gash DM, Gerhardt GA (2002) Effects of chronic intraputamenal infusion of glial cell line-derived neurotrophic factor (GDNF) in aged rhesus monkeys. *Neurobiol Aging* 23: 881–889

56   Nutt JG, Burchiel KJ, Comella CL, Jankovic J, Lang AE, Laws Jr ER, Lozano AM, Penn RD, Simpson Jr RK, Stacy M, Wooten GF (2003) Randomized, double-blind trial of glial cell line-derived neurotrophic factor (GDNF) in PD. *Neurology* 60: 69–73.

57   Kordower JH, Palfi S, Chen EY, Ma SY, Sendera T, Cochran EJ, Cochran EJ, Mufson EJ, Penn R, Goetz CG, Comella CD (1999) Clinicopathological findings following intraventricular glial-derived neurotrophic factor treatment in a patient with Parkinson's disease. *Ann Neurol* 46: 419–424

58   Yi A, Markesbery W, Zhang Z, Grondin R, Elseberry D, Gerhardt GA, Gash DM (2003) Intraputamenal infusion of GDNF in aged rhesus monkeys: distribution and dopaminergic effects. *J Comp Neurol* 461: 250–261

59   Brundin P (2002) GDNF treatment in Parkinson's disease: Time for controlled clinical trials? *Brain* 125: 1449–1451

60   Gill SS, Patel NK, Hotton GR, O'Sullivan K, McCarter R, Bunnage M, Brooks DJ, Svendsen CN, Heywood P (2003) Direct brain infusion of glial cell-line derived neurotrophic factor in Parkinson's disease. *Nat Med* 9: 589–595

61   Langston JW, Widner H, Goetz CG, Brooks D, Fahn S, Freeman T, Watts R (1992) Core assessment program for intracerebral transplantations (CAPIT). *Mov Disord* 7: 2–13

Progress in Drug Research, Vol. 61
(L. Prokai and K. Prokai-Tatrai, Eds.)
©2003 Birkhäuser Verlag, Basel (Switzerland)

Altering the properties of the blood-brain barrier: disruption and permeabilization

By David Fortin

Departments of Neurosurgery and
Neuro-oncology
University of Sherbrooke Hospital
Sherbrooke, Quebec J1H 5N4,
Canada

David Fortin

graduated from Sherbrooke University, Canada, in 1991 with an MD, and completed his training in neurosurgery in 1997 at the same institution. He then pursued fellowship training in neuro-oncology with Drs. Gregory Cairncross and David MacDonald at the University of Western Ontario, London, Canada (1998). His interest in neuro-oncology and in the limitation in therapeutics delivery imposed by the blood-brain barrier led him to actively pursue a fellowship training position with Dr. Edward Neuwelt and colleagues at Oregon Health Sciences University, Portland. He was appointed as Assistant Professor (1998–2000). He joined the Faculty of Medicine at Sherbrooke University and was appointed as the Neurosurgery Residency Training Program Director and Assistant Professor. He founded the neuro-oncology laboratory and opened an active clinical research program in neuro-oncology focused on blood-brain barrier manipulations. His research interests are glioma cell cultures and implantation models, in vivo assessment of therapeutics efficacy in glioma treatments, blood-brain barrier manipulations to improve therapeutics delivery, and in vivo growth factor targeting in gliomas. The Sherbrooke neuro-oncology clinical research program is an active part of the international blood-brain barrier consortium. More than 530 blood-brain barrier disruption procedures have been performed at this center since 2000.

Contents

Key words

Blood-brain barrier, brain tumor, disruption, osmotic opening, permeabilization.

Glossary of abbreviations

BAT, brain adjacent to tumor; BBB, blood-brain barrier; BBBD, blood-brain barrier disruption; BCNU, 1,3-bis(2-chloroethyl)-1-nitrosourea; BDT, brain distant to tumor; CNS, central nervous system; CSF, cerebrospinal fluid; CT, computed tomography; ICP, intracranial pressure; i.v., intravenous; LH, left hemisphere; MRI, magnetic resonance imaging; PET, positron emission tomography; SPECT, single photon emission computed tomography.

1   Introduction

By way of its intrinsic structural and physiological properties, the blood-brain barrier (BBB) represents a formidable obstacle to the delivery of drug to the central nervous system. The treatments of many diseases affecting the central nervous system is thereby complicated by the aspect of delivery, that is, insuring that the therapeutic molecule will reach the target cell in sufficient concentration, and in a suitable timing for the treatment to be effective. Although many different etiologic conditions will be affected by this delivery impediment, in no other condition has it been as extensively documented as in malignant brain tumors. Brain tumor is the prototypical situation through which one can best exemplify the problematic of delivery across the blood-brain barrier. We will therefore frequently refer to this particular problematic, acknowledging the fact that a lot of research endeavour in this field was undertaken as alternate strategies in the treatment of brain tumors. However, by no means will we imply that these strategies should be restricted to the treatment of cerebral malignancies.

Chemotherapeutic drug trials for brain tumors have been conducted worldwide for more than 4 decades, and most investigators agree that little progress has been made since the introduction of the nitrosoureas, a class of chemotherapy agents [1]. Moreover, the impact of these molecules on the natural course of malignant astrocytic neoplasms is rather limited. So limited, in fact, that many authors believe that their use might not be warranted [2]. Limited therapeutic success in the treatment of CNS neoplasia with chemotherapy is generally attributed to two factors: natural or acquired resistance to chemotherapy expressed by tumor cells, and delivery impediment related to the blood-brain barrier [3]. Many other neurological pathologies remain unresponsive to treatment (Alzheimer disease, strokes, amyotrophic lateral sclerosis, Hungtington disease and Gaucher disease) despite the fact that gene therapy using viral vectors might be available [4]. In most of these examples, therapeutic efficacy is limited by the inadequate penetration of the therapeutic agent to the targeted cells in the central nervous system [3, 5].

By way of its anatomic and physiological properties (extensively discussed elsewhere in this book), the normal blood-brain barrier prevents passage of ionized water-soluble compounds with a molecular weight greater than 180 Da [5]. As an example, most currently available effective chemotherapeutic agents have molecular weights between 200 Da and 1200 Da. Although the integrity of the barrier is often compromised within the tumor, this alteration in permeability is variable and dependent on tumor type and size [3]. Moreover, it is extremely heterogeneous in a given lesion. Although the BBB is frequently leaky in the center of malignant brain tumors, the well-vascularized actively proliferating edge of the tumor (the brain adjacent to tumor or BAT), has been shown to have variable and complex barrier integrity [6]. Therefore, by steeply reducing the concentration of intravenously administered chemotherapeutic agent at the periphery of the tumor, the phenomenon of sink effect is yet another mechanism that can contribute to chemotherapy failure in CNS neoplasm treatment [7]. Therefore, a strategy to increase dose intensity to the CNS must take into account the impediment imposed by the barrier, and somehow, bypass it [8].

Interestingly, despite the fact that this limitation in delivery imposed by the barrier is more and more acknowledged, this topic (of delivery) remains underdeveloped and under discussed in the field of neurosciences [9].

Different approaches have been advocated to improve delivery across the blood-brain barrier. These approaches can be broadly classified as either local, regional or global in their ability to circumvent the BBB. The local delivery methods are best exemplified by direct stereotactic inoculation of therapeutic vectors. The convection-enhanced delivery strategy, or clysis (see below), is considered regional in its ability to deliver a molecule to the CNS. Finally, the global delivery involve methods allowing delivery to the whole CNS, the best example being the use of chimeric molecules exploiting carrier systems in the BBB (discussed in the chapter by Suresh P. Vyas, this volume). The different strategies can also be stratified according to the rational on which they are derived: they are either neurosurgical-based (invasive), pharmacological-based (lipid carriers) or physiological-based (exploiting endogenous carrier system) [10]. In this chapter, we will focus the discussion on the transient osmotic permeabilization of the BBB, an invasive approach offering the potential of global delivery. However, a brief description of the other neurosurgically-based delivery methods follows to allow comparison.

2   Convection-enhancement of drug delivery

Bulk flow of the interstitial fluid occurs under normal conditions within the brain, under vasogenic edema, and following infusion of solutes directly into the brain parenchyma. Bobo et al. [11] describe an animal model of fluid convection (or bulk flow). Using a cat model, the authors demonstrated that if a pressure gradient can be maintained during infusion, the distribution of even high molecular weight proteins may be enhanced significantly when using high flow micro-infusion. With infusion rates of up to 4 µl/min, tracers were homogeneously distributed 1.5 to 2.0 cm from the infusion source. In a follow-up study, Lieberman et al. demonstrated in a series of four animal studies that convection-enhanced infusion can result in both greater concentration and a homogeneous distribution of larger molecules (molecular weight 10–126 kDa) in the infused region [12]. In a recent abstract, Rainov and colleagues report promising results of oncolytic adenovirus to treat recurrent glioblastoma using convection-enhanced delivery in a single patient [13].

Kroll et al. [14] replicated the results obtained in the cat models discussed above, in rats. These authors assessed which of three factors (dose, volume and infusion time) would be most influential in increasing the volume of virus-sized superparamagnetic particles to the brain. The total dose was the most important of these factors suggesting convection may not be as important as the total dose administered. Along with an increase in dose, a slower infusion time also led to higher tissue volume. Thus, the ability to increase dose at a slow delivery rate may lead to higher doses of chemotherapy within the brain. Clinical studies are underway, evaluating this novel strategy. Muldoon et al. demonstrated the convection-enhanced delivery of virus particles and paramagnetic iron oxide nanoparticles using intracerebral inoculation vs. osmotic disruption of the BBB [15]. The former approach demonstrated local delivery of the viruses while blood-brain barrier disruption (BBBD) demonstrated more global distribution of the virus particles to the disrupted hemisphere.

3　Increase in vascular permeability

RMP-7 is a bradykinin B2 receptor agonist that has been shown to selectively increase permeability of the vasculature supplying brain tumors in both animal models and humans. The increase in permeability produced by this agent occurs rapidly but is very short, lasting only 2 to 5 minutes following cessation of infusion. Even with continuous infusion, spontaneous restoration of the barrier begins to occur within 10 to 20 minutes, limiting the therapeutic window required to deliver medication. Also, as this approach seems to be more effective for the blood-tumor barrier than for the blood-brain barrier, this approach may hold better promise for brain metastases than for primary brain tumors, or for other non-neoplastic application. The delivery system overall is well tolerated with very few related adverse events [16]. Thus far, it has been tested mostly with carboplatin in phase I studies and in dose escalation.

4   Local delivery

This approach involves the local application of the therapeutic vector, whether it be "wafers", containing a fixed concentration of chemotherapy, stem cells, antibody conjugate or gene producing cells. Thus far, wafers have been tested with 1,3-bis(2-chloroethyl)-1-nitrosourea (BCNU) and are now commercially available for use in recurrent gliomas [17]. Other drugs are being tested using the same strategy [18]. Although conceptually advantageous with respect to the low systemic exposure to the therapeutic agent, this approach has one notable limitation: the poor diffusion of the molecule in the brain parenchyma resulting in uneven drug distribution. The same restriction would apply in the treatment of other global CNS pathologies.

5   Blood-brain barrier disruption (BBBD)

This approach involves the cerebral intravascular infusion of hypertonic solutions to produce a transient increase in permeabilization of the barrier, in a given cerebral distribution (carotid or vertebral) (Fig. 1). There is now extensive animal and human clinical data on the use of this approach [3, 5, 7].

When one realizes the extensiveness of the vascular network supplying the brain, it becomes obvious that a global delivery strategy is plausible by using this vascular network as a vector of delivery. The importance of this network has already been exposed by Bradbury, stating that the entire network covers an area of 12 m²/g of cerebral parenchyma [19]. In fact, the extensiveness of this network is further exemplified by the fact that the brain receives 20% of the total cardiac output, even though its weight constitutes only 2% of the total body mass [20]. By increasing the permeability of these vessels, inactivating the function of the BBB, the perfect mean of global delivery is at hand. This inactivation in the BBB function obviously needs to be transitory, but of sufficient time to allow the intra-arterial infusion of a therapeutic molecule. It is precisely what the BBBD procedure offers.

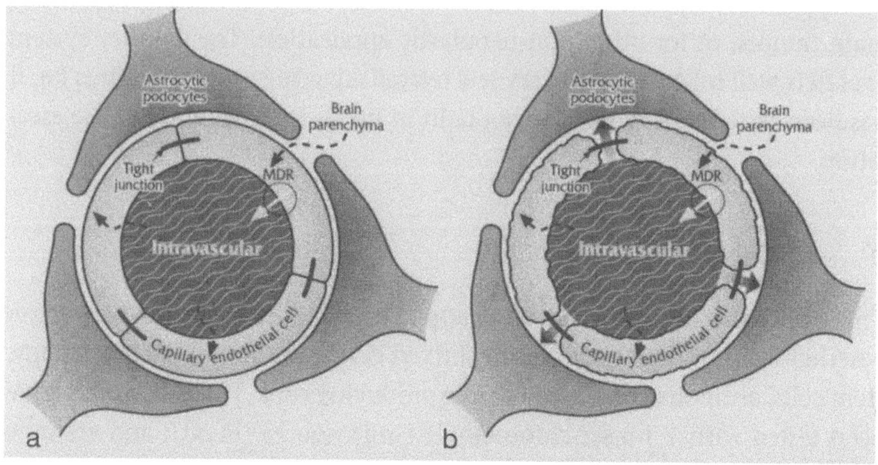

Figure 1.
Graphical sketch illustrating the hypothesis concerning the osmotic BBB modification. The tight junctions are shown (a) as devoid of any anatomic space between the endothelial cells. Moreover, the multi-drug resistance gene product (MDR), or p-gp efflux pump is also illustrated as it is integral to the mechanism of the barrier. The osmotic BBBD procedure induces a retraction in the cell membrane, and a physical opening (b) accompanied by a modification of the Ca^{2+} metabolism in the cell.

5.1   Pre-clinical data

In the late 40's, Broman and Olson first reported on the effects of some angiographic contrast media on the permeability of cerebral blood vessels [21]. Twenty years later, Rapaport et al. described the effect of intra-arterial infusion of hyperosmolar solutions on the cerebral vasculature and first theorized on the osmotic opening of the blood-brain barrier [22]. Using intravascular Evans blue-albumin extravasation as a marker of permeability, these authors determined the different parameters consistent with a reversible opening of the BBB. Evans blue binds tightly but reversibly to albumin *in vivo*, resulting in a 68,500 Da molecular weight marker that normally does not cross the BBB. It was thus established that the blood-brain barrier can be disrupted after the intravascular infusion of a 1.4 M (1.6 molal solution). The osmotic threshold (the minimal osmolal concentration) of the hypertonic solution required to induce reversible BBBD varies with the lipid solubility of the solute as measured by the octanol/water partition coefficient [7]. The osmolal threshold increases with increasing lipid solubility. Hypertonic solutions used to disrupt the BBB include mannitol, arabinose, lactamide, saline, urea and several radiographic contrast agents. Because it is approved for administration to patients, mannitol has become the agent of choice in both preclinical and clinical studies [5]. Opened tight junctions were visualized using electron microscopy with the tracer horseradish peroxidase after intracarotid infusion of hypertonic solutions in different animal species [23]. Moreover, using an *in vitro* preparation of monolayer endothelial cells as a model of the BBB, Dorovini-zis et al. were able to show the opened junctions after exposition to a hypertonic solution [24].

A number of substances and physiological alterations have also been studied for their ability to modify the properties of the barrier. These include pentylenetetrazol, dimethyl sulfoxide, etoposide, 5-FU, vineralbin, hypertension and hypercapnia. For reasons such as inconsistency of barrier opening, duration of barrier opening, resultant structural damage and irreversibility, none of these agents is likely to be clinically applicable [5].

5.2 Parameters involved in the BBBD

Two parameters are paramount in the ability to mediate a hyperosmolar modification of the barrier: the osmolality of the solution, and the infusion time. Using a solution of 1.6 molal arabinose in pentobarbital-anesthetized rats, Rapaport determined an interval of 30 seconds as the optimal infusion time for BBBD [25]. The same infusion time was applied to the use of mannitol with similar findings in the same animal model [7]. Our laboratory has identified 22 seconds as a minimum infusion time to produce BBBD in the Long-Evans rat model. However, this was produced at the expense of a higher infusion rate with an increased rate of hemorrhagic complication (submitted). Therefore, the 30 seconds infusion interval is considered standard by most authors.

BBBD requires general anesthesia for a number of reasons [3, 5, 7]:

1. The procedure generates a significant level of pain.
2. It causes a transient rise in intracranial pressure requiring cerebral protection. An actual increase in brain water (vasogenic edema) of 1–1.5% has been documented after the procedure.
3. The mannitol infusion induces hemodynamic instability. This is illustrated in the treated animals by a brief (20 to 30 seconds) period of apnea, hypotension and bradycardia [7]. The same phenomenon is observed in

the humans undergoing the procedure, and must be pharmacologically prevented (see section on the clinical procedure).

The choice of anesthetic agent has proven important in the quality of BBBD, as well as the potential for toxicity related to the drug treatments [26, 27].

Ketamine, as well as other anesthetic agents used for BBBD, was found to be inconsistent, producing only 40-70% of good to excellent BBBD [26]. It is presumed that negative impact on the cardiac index, as well as on cardiac rate and systemic arterial pressure reduces the effectiveness of the procedure by altering cerebral blood flow, thus decreasing the effective rate of mannitol delivery during the intra-arterial infusion [26]. Relative inconsistencies in BBBD obtained when using certain anesthetic agents prompted efforts to modify the model. Propofol has been shown to be the most efficient agent, producing greater than 95% of good to excellent BBBD in animals (Fig. 2) [27]. With increase in both the consistency and the intensity of the BBBD, propofol has also been found to produce neuro-toxicity with certain chemotherapy agents not previously reported as toxic with other anesthetics [28]. Hence the choice of a specific anesthetic agent may be accompanied by undesirable toxicity eliminating any potential advantages.

5.3 Monitoring the degree of BBBD in animal models

Traditionally, a 2% solution of Evans blue is administered intravenously (2 ml/kg of body weight) in the animal prior to the BBBD procedure [7]. As mentioned earlier, the marker binds tightly but reversibly to albumin *in vivo* resulting in a 68,500 Da molecular weight marker that does not cross the intact BBB. It thus provides a semi quantitative visual measure of osmotic BBB modification (Fig. 3, Tab. 1). Another approach has been described in which a perfusate containing ^{14}C-sucrose is infused intra-arterially post disruption. The brain is then harvested, digested and scintillation counting is accomplished [29].

Our laboratory is currently developing a quantitative approach, using immunochemistry to label albumin (Fig. 4). The slides are then scanned and treated in an imaging software developed by Le Dinh and his team (Sherbrooke University). The images are converted in gray scales matrix. A numer-

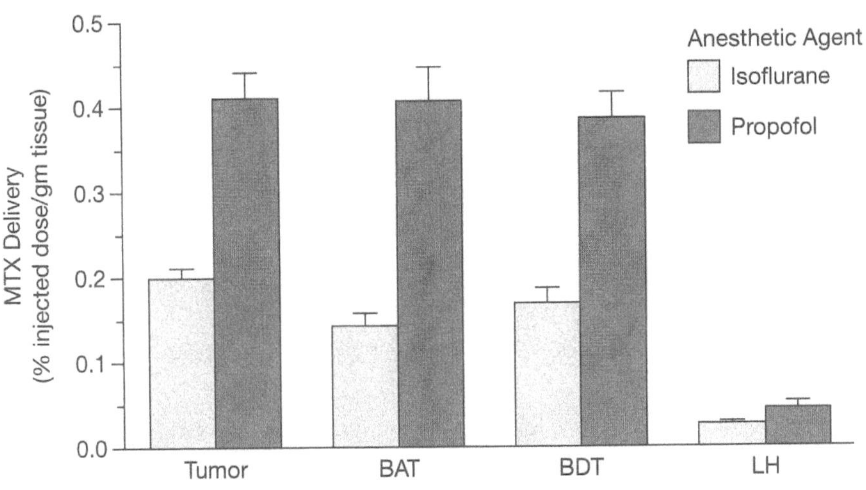

Figure 2.
Graphic depicting the difference in delivery obtained by modification of the anesthetic agent. The comparison was carried between isoflurane and propofol for ^{3}H-methotrexate. The graphic is broken down according to the regional assessment of delivery for the tumor, brain around tumor (BAT), brain distant to tumor (BDT) and undisrupted control hemisphere (LH). (Results obtained in Dr. Neuwelt's laboratory.)

ical value is assigned to the number of pixels above a standard threshold measuring the immunostaining, and this value is assigned as the degree of BBB modification. The disrupted hemisphere value assigned is normalized using the contralateral hemisphere for each animal. This approach has been found to highly correlate Evans Blue staining, while providing a more precise measurement and an objective value (submitted). It has the notorious advantage of allowing the study of the topographic distribution of the BBB modification (Figs. 4 and 5).

Other techniques have been applied to measure the degree of BBBD in the animals, using computed tomography (CT) scan, magnetic resonance imaging (MRI) and radionuclide imaging [5]. While providing antemortem data on the degree of disruption in the treated hemisphere, these techniques are cumbersome and expensive, and cannot be used routinely. We are about to initiate a positron emission tomography (PET) scan study using coopermarked albumin that will allow us to perform antemortem assessment of BBBD.

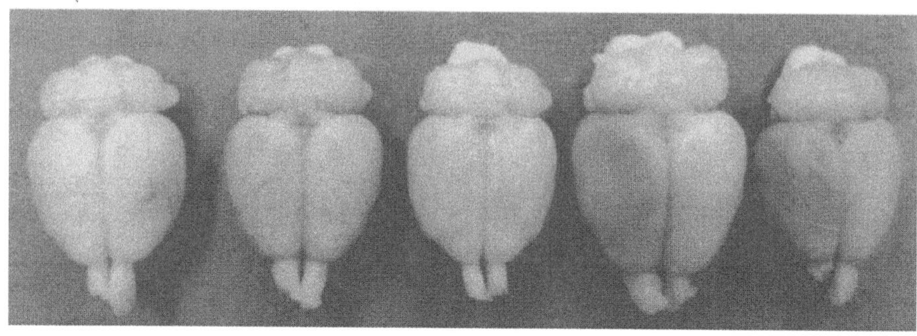

Figure 3.
Sequential discoloration of the right hemisphere in a series of Fischer rats infused with intravenous (i.v.) Evans blue and treated with BBBD according to our modified model. 0.12 cc/sec has been identified as the optimal infusion rate.

Table 1.
Grading scale for descriptive quantification of Blood-Brain Barrier Disruption (BBBD) provided by Evan's blue staining

BBBD grade	Description
Grade 0	No blue staining of the cerebral parenchyma
Grade I	Slight blue tint to the cerebral parenchyma in the territory supplied by the parent artery infused with the mannitol
Grade II	Clearly demarcated blue staining of the cerebral parenchyma in the territory supplied by the parent artery infused with the mannitol
Grade III	Blue staining of the cerebral parenchyma which tends to surpass the territory supplied by the parent artery infused with the mannitol *via* the polygon of Willis
Grade IV	Extreme blue staining of the cerebral parenchyma which surpasses the vascular territory infused with the mannitol

5.4   Animal models

Different animal BBBD models have been developed, but the rat, for its ease of use, is the most frequently cited [30]. Studies in normal rats demonstrated that greater than 95% of the animals attained good or excellent BBBD when infused with 1.4 mol/L mannitol when using propofol. In contrast, in rats with brain tumor xenografts, only approximately 60% of animals had good or excellent disruption [27].

The standard technique involves the surgical exposition of the right

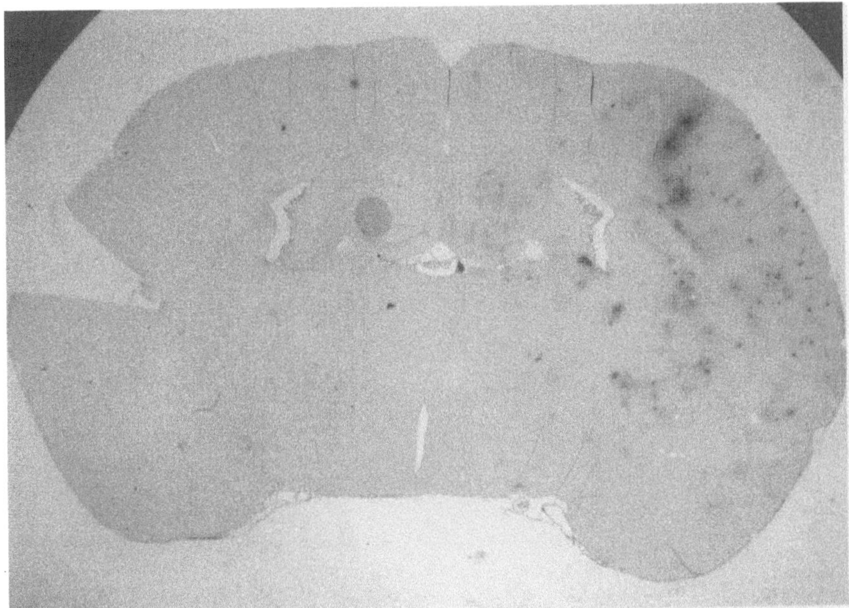
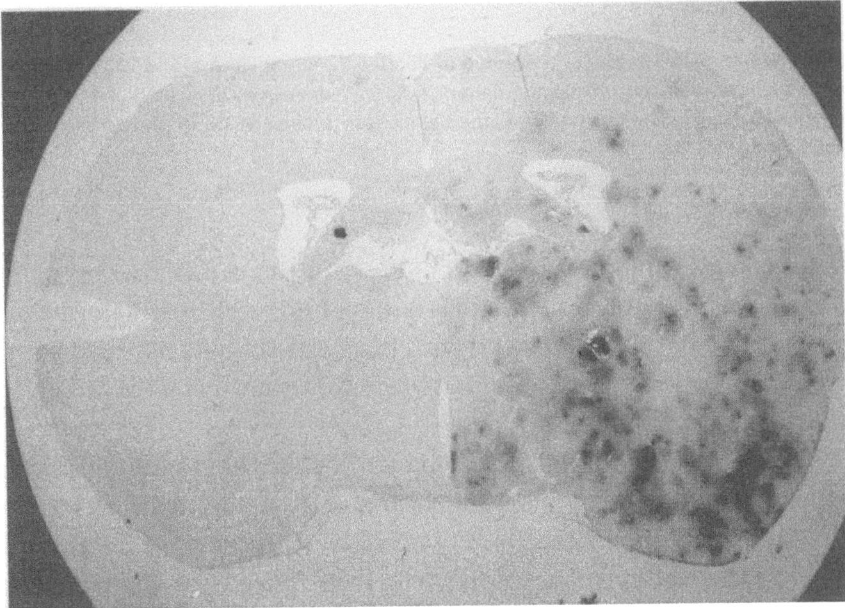

Figure 4.
Coronal samples of non-tumor bearing Fischer rat brain exposed to BBBD with our modified model. An albumin immunohistochemistry was accomplished, and is illustrated by the black discoloration in the right hemisphere. The mannitol infusion rate was 0.08 cc/sec in (a) and 0.12 cc/sec in (b). The difference in staining intensity can be appreciated, as well as the topographic heterogeneity in capillary leaking.

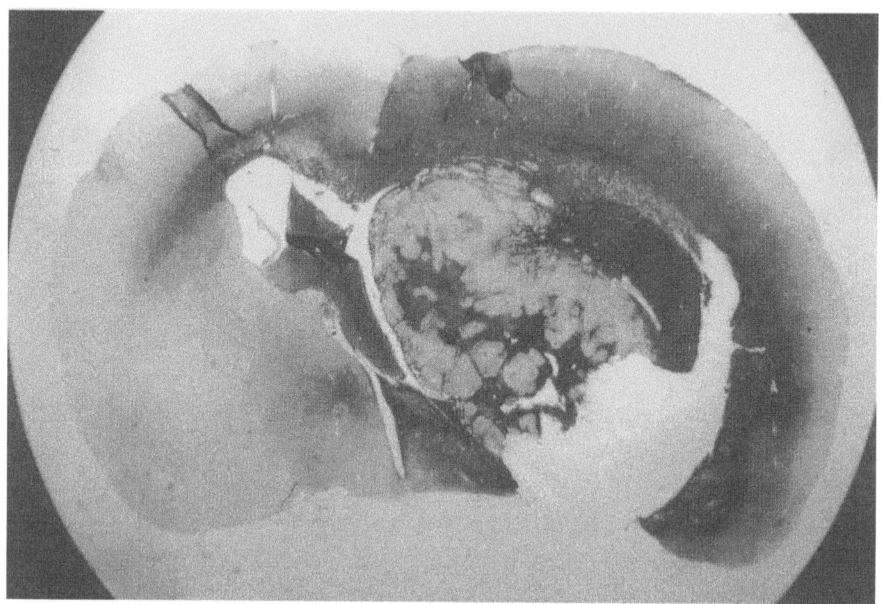

Figure 5.
Coronal sample of Fischer rat brain exposed to BBBD with our modified model 20 days after the implantation of a F98 tumor. An albumin immunohistochemistry was accomplished, and is illustrated by the black discoloration illustrating the marked increase in permeability in the right hemisphere around the tumor.

carotid complex under aseptic technique, and the catheterization of the external carotid artery in a retrograde fashion using a PE-50 intramedic tubing so that the tip of the catheter is lying just above the bifurcation. The solution of 25% mannitol is then administered intra-arterially *via* the catheter in the external carotid artery.

We have modified the model to improve the effectiveness and consistency of the BBBD procedure while using ketamine as an anesthetic agent, thus decreasing the risks of neurotoxicity observed with the pre-clinical use of propofol. The modification simply involves the placement of a temporary vascular clip to the common carotid artery approximately 1 cm proximal to its bifurcation prior to the infusion of mannitol (Fig. 6). This simple step allows the isolation of the perfused hemisphere from the hemodynamic effects induced by the anesthetic agent upon the cardiovascular system. It simplifies the hemodynamic system by decreasing the number of variables

Figure 6.
Surgical dissection of the right carotid artery complex in the rat, demonstrating the application of a temporary vascular clip (black arrow head) to the right common carotid artery (white arrow head). A PE-50 intramedic catheter (double arrow) cannulates the external carotid artery allowing infusion of mannitol in a retrograde fashion into the internal carotid artery (single arrow) to produce osmotic blood-brain barrier disruption.

involved. The rate and duration of infusion then become the sole relevant modifiable parameters.

It is of note that the parameters considered optimal for the BBBD procedure are different and must be adjusted for the strain of rat, the size of the animals and the anesthetic used. Reviewing the BBBD approach in seven experimental brain tumor models, Blasberg et al. concluded that vascular permeability varies tremendously among models, and the threshold for BBBD within tumor is directly related to permeability of the tumor before disruption [30].

5.5   From pre-clinical data to the clinic

From the animal studies, critical information was obtained and later correlated in the human situation [3, 5, 7]. These data were paramount in the deployment of a standardized BBBD procedure in the clinic.

The procedure has been found to produce a marked increase (10- to 100-fold) in brain and cerebrospinal fluid (CSF) concentrations of methotrexate and of other markers [7] (Fig. 7). The alteration in permeability is variable within different brain regions and within different types and sizes of tumors [30].

Typically, the ipsilateral cerebral cortex, containing the highest density of capillaries, is the area where the permeability is mostly increased by the BBBD procedure. Interestingly, this regional heterogeneity favoring the cortical area accounts for the fact that the procedure evades the problem of the sink effect, the rapid equilibration into normal brain of drug delivered primarily to tumor in sub therapeutic concentrations, by providing higher and more uniform delivery to the whole CNS [5]. This decreases the rate at which molecules diffuses away from the tumor, resulting in prolonged tumor exposure to a higher concentration of drug [31, 32].

Another interesting finding derived from the pre-clinical studies was the impact produced by radiotherapy on delivery. The administration of external beam radiation therapy either prior to or concurrent with the administration of a high molecular weight marker ^{14}C-labeled dextran 70, or a low molecular weight marker ^{3}H-labeled MTX resulted in a statistically significant (P < 0.01) decrease in drug delivery when compared to animals not receiving cranial irradiation in a rat model [33]. Therefore, previous cranial irradiation decreases the delivery of agents to brain after osmotic BBBD. Radiotherapy is considered a standard treatment for malignant brain tumors and is regularly used, even if it has the potential to induce significant neurotoxicity. The fact that it also further decreases drug delivery after BBBD should be considered when evaluating the efficacy of BBBD in patients previously irradiated. Ideally, radiotherapy should not be administered before BBBD.

Neurotoxicity, however, might also be produced by increasing the delivery of molecules that are normally restricted in their CNS penetration. This has been exemplified in a number of studies. Severe neurotoxicity has been documented when using BBBD in animal models to increase the delivery of chemotherapy agents such as cisplatin and doxorubicin [5]. While modifying the animal model of BBBD by using propofol as the anesthetic agent with the goal of further improving delivery, Fortin et al. demonstrated an increase in toxicity with agents that were never identified as toxic before the use of this modification [28]. That is, agents routinely used without toxicity in the animal and the humans in the context of BBBD enhanced delivery became toxic when the enhanced delivery was further maximized. This illustrates the fact

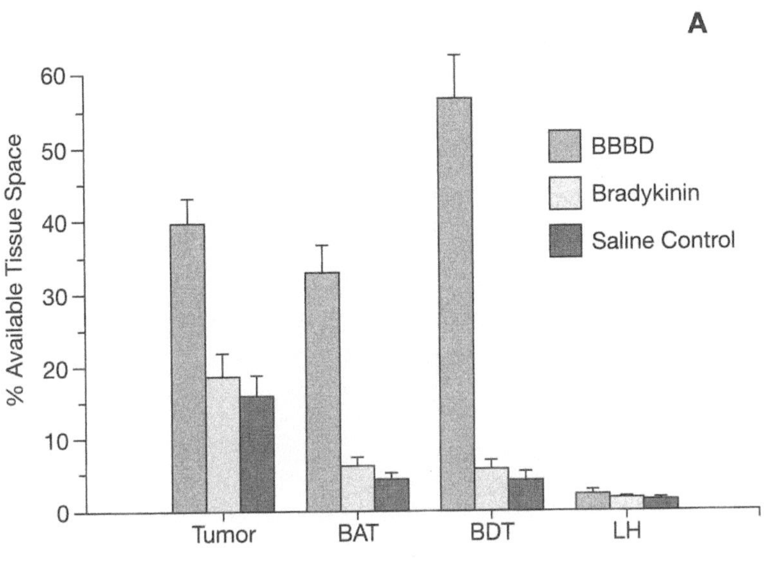

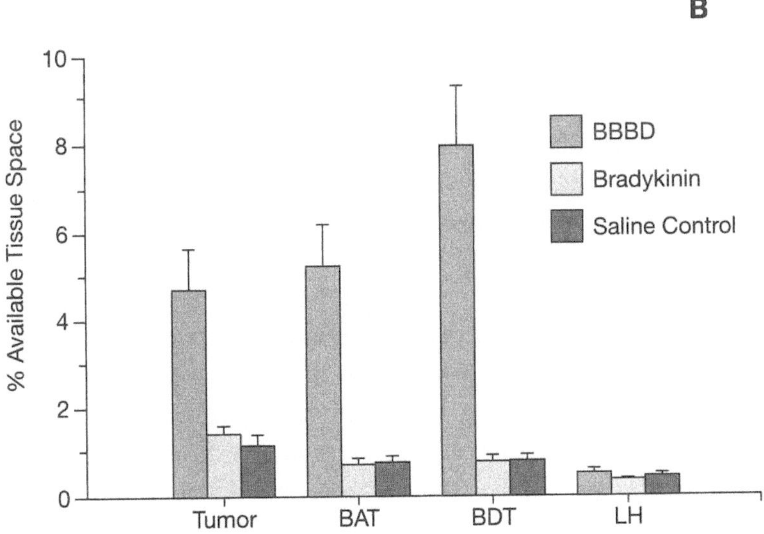

Figure 7.
Graphic illustrating the increased delivery of r   te (A), and of Dextran (B) in the tumor, brain around tumor (BAT), brain distant to tumor (BDT) and the undisrupted left hemisphere (LH). The graphic depict the comparison in delivery after blood-brain barrier osmotic disruption, bradykinin infusion, and a saline control infusion. (Results obtained in Dr. Neuwelt's laboratory.)

that potential complication related to this strategy should not be taken lightly. That being said, the clinical experience so far has demonstrated the safety of this procedure, and as will be discussed in the following section, the procedure is now well standardized and its application in the human is already a reality with objective data on its potency to increase delivery to the CNS.

Different molecules have been used in conjunction to the BBBD procedure in the animal models. Different chemotherapy agents, monoclonal antibodies, superparamagnetic particles and virus have been successfully delivered with this approach [35–40]. In the human, thus far, the experience has been limited to the use of chemotherapy agents, and of boronophenylalanine in the context of brain tumor treatment [41, 42].

5.6 The clinical experience so far

Neuwelt and his group conducted the first phase I clinical studies on osmotic opening of the blood-brain barrier, beginning in 1979 [3, 7]. After a series of phase I studies (toxicity assessment) had been successfully completed, phase II studies (efficacy assessment) were initiated, and are still underway, paving the way for phase III studies (randomized study assessing the efficacy of the procedure against a treatment considered as standard). Using a standard technique of osmotic blood-brain barrier disruption to enhance chemotherapy delivery with three different chemotherapy protocols, more than 3000 procedures have been performed in more than 300 patients across the BBBD consortium, an entity which includes six university centers coordinated by the Oregon Health Sciences University [41].

The data derived from this consortium have been reported [41]. As mentioned previously, the experience has been limited to brain tumor treatment, and since this is not the focus of this chapter, the results will not be discussed in detail. However, this study is worthy of our attention because it basically demonstrates for the first time that by using a standardized protocol, the BBBD procedure is safely applicable in the setting of a multicenter effort. Dr. Neuwelt and his team must be entirely credited for pioneering this clinical approach.

5.7 Description of the clinical BBBD procedure

Since general anesthesia is required, the osmotic opening of the blood brain barrier is accomplished in the operating room or in the angiography suite

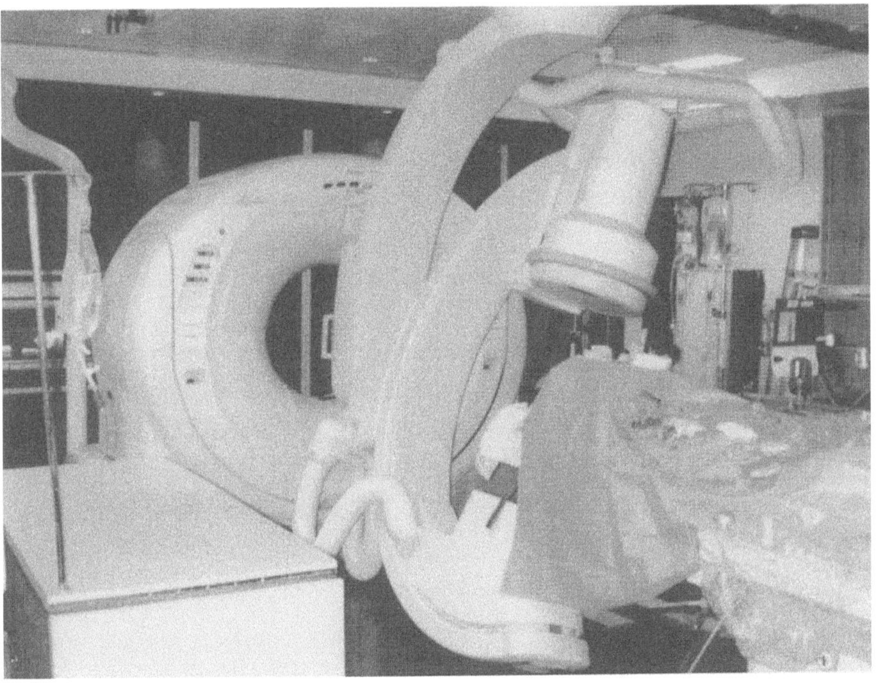

The Sherbrooke University Hospital center angio-ct suite, allowing the acquisition of rapid sequential CT-scan during and after the BBBD.

(Fig. 8). The anesthetic agent used for the procedure has traditionally been isoflurane, but the BBB international consortium is evaluating the possibility to use other agents to maximize the consistency of the procedure.

The human cerebral arterial system is organized in such a way that there are basically 4 major arteries responsible for the brain irrigation. The vascular anatomy can be variable from one individual to another, and thus the precise anatomy must be determined during the first treatment session by a formal cerebral angiography. The nature of the procedure, by producing a vasogenic edema in the treated vascular distribution, do not allow for disruption of more than one vascular territory in a single treatment session. This implies that if a lesion covers more than one vascular distribution, different treatment session will have to be offered to the patient to complete one cycle of treatment.

As hemodynamic factors impact on the degree of disruption, every attempt is made to keep heart rate and systemic pressure stable and above threshold values (established for each patient) during the general anesthesia [3, 41].

After general anesthesia, the technique involves the following steps:

1. Selective catheterization *via* percutaneous transfemoral puncture of left internal carotid artery, right internal carotid artery, left vertebral artery and right vertebral artery. The tip of the catheter is positioned at the C2-C3 vertebral level in the carotid, or at the C6-C7 vertebral level in the vertebral artery.

2. Determination of rate of infusion of mannitol by iodinated contrast injection and fluoroscopy, as the lowest infusion rate in which there is retrograde flow from the arterial catheter. The volume of mannitol infused will be the rate determined in cc/sec × 30 sec (usually between 4 to 12 cc/sec in carotid circulation, and between 4 to 10 cc/sec in vertebral circulation). The ultimate goal is to fill the entire vascular compartment in a given vessel distribution, without producing backflow of mannitol in the parent vessel.

3. The osmotic disruption is a physiologically stressful procedure. It can induce focal seizures in 5% of procedures [41]. It can also trigger a vasovagal response with bradycardia and hypotension. In order to prevent the occurrence of these adverse effects, the following medications are administered just prior to the disruption:
   A. Diazepam 0.2 mg/kg IV (maximum dose = 10 mg).
   B. Thiopental 1 to 3 mg/kg IV.
   C. Atropine IV, titrated to increase heart rate 10 to 20% from baseline (0.5 to 1 mg).
   D. Ephedrine, if needed, to obtain systolic blood pressure of at least 100 mm HG.

   Moreover, as mentioned earlier, the procedure induces a low to moderate transient rise in intra-cranial pressure. To minimize this effect, the patient is hyperventilated 3 to 5 minutes with 100% O_2 to obtain an end tidal CO_2 of 30 to 35.

4. Osmotic disruption of the blood brain barrier is accomplished by infusing mannitol, 25%, in the previously catheterized artery at the previously defined rate.
   - During the infusion, interesting signs can be observed. The medial aspect of the forefront ipsilateral to the side of the infusion undergoes

a whitish discoloration explained by the washout of blood from the ethmoidal branches, arteries known to connect the intracranial circulation to the extra cranial circulation. Bilateral pupillary dilatation can also be observed during the mannitol infusion. This is followed by a brief period of tachycardia and systemic hypertension.

5. Intra-arterial contrast infusion to confirm catheter position and rule out arterial injury post-disruption.

6. Infusion of the therapeutic molecule intra-arterially in the disrupted circulation. The concentration of the solution and the rate of infusion are critical factors when infusing intra-arterial solutions in avoiding neurotoxicity. The phenomenon of streaming defines an inhomogeneous distribution of the administered solution because of poor mixing at the infusion site [43]. It is directly related to the Reynold number, a crucial parameter in fluid dynamics that predicts the transition from streamlined to turbulent flow [43]. In the equation leading to the Reynold number, the density and viscosity of fluid, lumen diameter of the infused vessels and velocity of flow are all important determinant to control in order to avoid streaming [44].

7. Termination of procedure. The patient is taken to recovery room, where his neurological status and state of consciousness will regularly be evaluated until full recovery.

If the degree of BBBD opening needs to be assessed, a standard dose of intravenous (i.v.) non-ionic contrast bolus can be administered five minutes after the mannitol infusion. At the end of the procedure, before proceeding to the recovery room, the patient is taken to the CT scan suite. The non-ionic contrast is a water-soluble agent, and its distribution across the barrier is therefore a good reflection of the extent of the barrier opening (Fig. 9). The contrast enhancement in the distribution of the disrupted vessel is linear with the degree of disruption. A visual grading system described by Roman-Goldstein et al. can be used to evaluate the degree of disruption [45].

The classification used is as follows:

1   Nil (no disruption)
2   Moderate
3   Good
4   Excellent

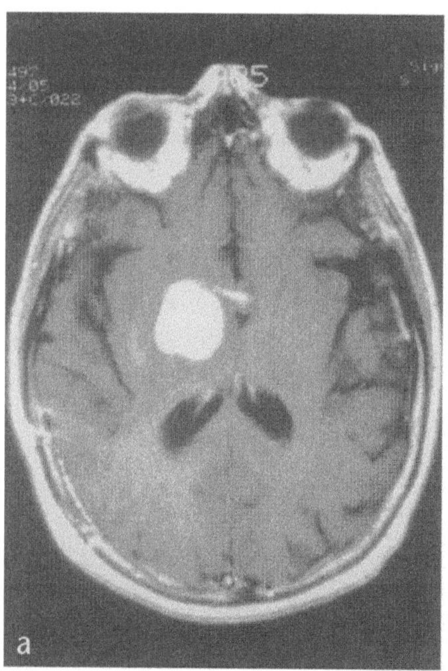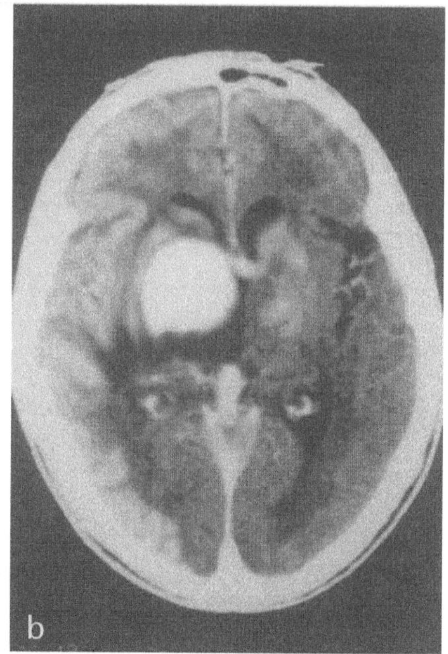

Figure 9.
(a) Pre-treatment MRI with gadolinium enhancement in the right basal ganglia of this 67 y.o. patient. A central nervous system lymphoma was diagnosed by a biopsy. (b) An iodinated contrast-enhanced CT scan in the same patient 20 minutes after BBBD in the right hemisphere. Beside the uptake of iodinated contrast material in the right hemisphere, notice the area around the tumor that was not uptaking gadolinium contrast on (a), that now brightly enhances. This area is considered the BAT (brain around tumor), and now is permeable, as well as the brain distant to tumor in the right hemisphere to the therapeutic molecule that will be administered (methotrexate, in this case).

Grade 3 BBBD (good) is generally considered the ideal degree of permeabilization. Grade 4 disruptions (excellent) tend to be paralleled by transitory neurological deficits and more prolonged unconsciousness, and are therefore not advisable for most patients [41]. On the other hand, grade 1 disruptions are probably not significantly effective in increasing delivery across the BBB. Grade 2 might be acceptable, but grade 3 should be aimed for. This visual grading has not been directly correlated with the factual increment in drug delivery. However, and interestingly, when reviewing our series of patients with primary CNS lymphoma treated with BBBD and methotrexate tri-drug regimen, our data did show a relationship between the degree of the disruption, as assessed by contrast enhanced CT scan, and survival in those

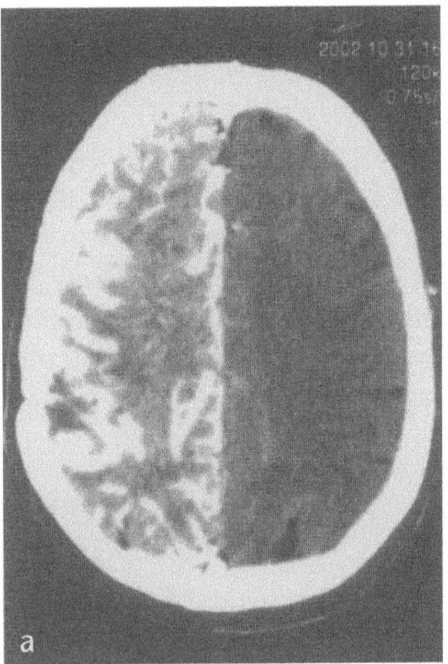
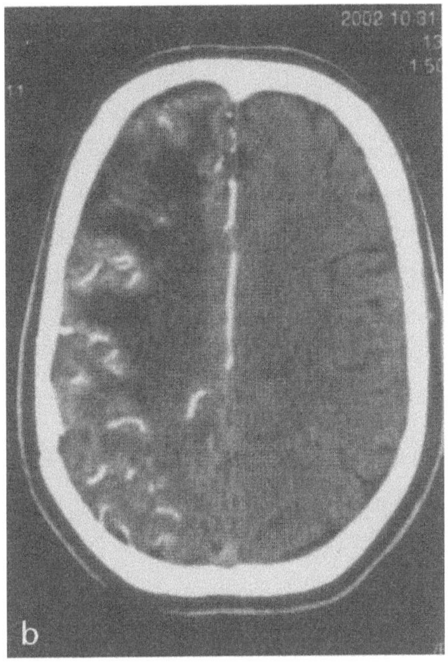

Figure 10.
(a) Serial CT scan obtained 5 minutes after BBBD, and (b) 10 minutes after BBBD. The iodinated contrast product was infused 1 minute after the mannitol infusion. Note the important contrast stagnation after 5 minutes, and the impressive decrease after 10 minutes. The contrast is illustrated as a whitish signal in the brain parenchyma.

patients [46]. It therefore clearly established a relation between the clinical outcome of the patient and the intensity of delivery obtained by the procedure.

We are currently investigating the dynamic of the vascular alteration in permeability using sequencing CT scan imaging during and after the intrarterial mannitol infusion in the context of a pilot clinical study (Fig. 10) with the goal of developing a more precise measurement tool for the degree and duration of the disruption. Although it is well accepted that the barrier is maximally opened for 20 minutes, some studies have demonstrated an opening for as long as 8 hours in some patients [47]. Using SPECT study in 12 patients, Siegal and collegues found that the barrier was open during the first 40 minutes after the procedure and returns to a functional baseline levels only after 6 to 8 hours following the induction of good or excellent disruption [47].

5.8   Contraindications to the BBBD procedure

As stated earlier, the blood-brain barrier disruption procedure induces a transient rise in intra-cranial pressure (ICP; baselines 3 to 9 cm H_2O to peaks of 16–23 cm at 30 min post disruption). This transient rise in ICP has been shown to correlate with a 1.5% increase in brain fluid content [7,48]. In pre-clinical studies, this transient increase in ICP was not associated with any clinical sequelae [7]. It illustrates however the rationale behind the very first contraindication of osmotic BBBD: the presence of a significant mass effect. The definition of mass effect is somewhat arbitrary. Therefore, the following criteria must be used in clinic, to provide some standardization:

1   Patients with radiological signs of herniation are ineligible.
2   Patients with any radiological signs of compromise of the basal cisterns are ineligible.

All other patients displaying significant radiological signs of mass effect, but not strictly adhering to these two absolute contraindications must be individually and carefully screened.

Other contraindications include:

1   Patients with evidence of spinal cord block from tumor mass.
2   Patients at significant increased risk for general anesthesia.
3   Patients that underwent previous vascular surgeries (including but not limited to femoral graft, carotid endarterectomy).

In the clinical studies underway, other contraindications were stipulated directly referring to the chemotherapeutic agents, and are therefore not mentioned here.

5.9   Adverse effects

In a seminal paper published by McAllister et al., the effect of the procedure combined to chemotherapy infusion (methotrexate) on neurocognitive function was evaluated in long term survivors from the procedure. Contrary to numerous established treatment modalities (e.g. cerebral radiotherapy), the BBBD procedure was not associated with any decline in formal neurocognitive assessment [49].

Adverse effects can nevertheless occur, and for ease of discussion, are stratified in two groups:

1 Catheter related complications
   - Asymptomatic subintimal tear during catheterization (incidence 5%)
   - Significant groin hematoma post catheterization (incidence 0.5%)
   - Parent vessel thrombosis, as experienced in two of our patients (incidence 0.5%). This complication can produce long term neurological disability related to the vascular distribution involved. Fortunately, the occurrence of this complication did not translate in clinical repercussion in our patients.
2 Disruption related complications
   - Seizures (incidence 5%). Seizures are typically focal, and are immediately treated with IV thiopental and/or IV Diazepam. This is typically a procedure related event.
   - Temporary obtundation and/or increase in neurological symptoms (incidence 2.5%). Complete recovery is the norm, and it typically lasts less than 48 hours. This adverse effect is associated with an excellent disruption.
   - Brain herniation. Obviously, the most serious complication. One patient expired from this complication (incidence 0.3%). It is the only mortality event related to the procedure in 300 patients that underwent a total of more than 3000 procedures [41].

5.10 Future developments

The literature supporting the use of the BBBD approach as a successful mean of delivery is considerable. It is one of the most studied delivery strategies to the CNS both pre-clinically and in the clinic. However, its use is still extremely limited. Only a few centers in the world treat patients with the osmotic BBBD procedure. This is explained by many factors: the procedure has a steep learning curve, it requires access to technical facilities on a regular basis, and it requires the presence of different specialists, all working on a weekly schedule as a team. The procedure is deemed "invasive" or "aggressive", even though the security and feasibility has been reported on, and tested in more than 400 patients in the context of multi-centers studies. Clinical research is underway to refine treatment of brain tumors with this approach. New chemotherapy agents are to be tested, in this notoriously chemo-resistant disease. New chemoprotectant agents are also to be tested, to decrease the systemic side-effects of chemotherapy. The barrier imposed by the BBB brings the opportunity to treat all CNS disease with the ideology of a two compartment disease, where the CNS dose of the therapeutic agent is maximized, while the systemic exposure is minimized or even nullified. This can be accomplished with the use of agents that would not cross the BBB, yet that would inactivate the systemic drug in circulation. These agents could be administered once the barrier is closed, therefore supporting the relevance of establishing more precisely the window of opening following mannitol infusion in each patient. One clinical example of this strategy is already used across the international BBB consortium with the administration of sodium thiosulfate as a rescue agent to the toxicity of carboplatinum [50, 51].

Pre-clinical research aiming at combining the osmotic BBBD procedure with the administration of other type of molecules is also actively underway. Viral vector, monoclonal antibodies and oligo-antisense are but a few examples of non-conventional agents tested in conjunction with the BBBD procedure. In face of the increasing and encouraging pre-clinical data, the first phase I studies in the humans combining BBBD to non-conventional molecule infusion should see the light in a few years.

6   Conclusion

In a recent editorial, Pardridge described his astonishment at the fact that the delivery issue is not treated with the proper respect it deserves when considering treatment of various central nervous system disease [9]. Amazingly, in the field of brain tumors, an important number of clinical studies are still underway with intra-venous administered agents possessing a poor penetration of the BBB. The scientific community involved in the field of neurosciences therapeutics must realize and acknowledge the role played by the barrier in delivery impediment to the CNS. Beside efficacy of a molecule for its target, the molecule must reach the target in sufficient concentration and time.

The BBBD procedure represents a concerted effort to produce a rational approach to this delivery impediment. It has shown its safety and its efficacy in increasing the delivery [41]. Its potency to modify clinical outcome in certain neoplastic disease has been established [46]. Other strategies are also under investigation. With continuous effort from groups dedicating their research attention toward this interesting problem, the neuroscience field will continue to accumulate data that will eventually allow more and more translational use of these strategies in the clinic.

Acknowledgements

I would like to express my kindest gratitude to Dr. Edward Neuwelt, a true pioneer in the field of CNS delivery. This work is as much a review on the interesting topic of delivery across the blood-brain barrier by osmotic means as a homage to his work in this field.

References

1   Fine HA, Dear KBG, Loeffler JS, Black PM, Canellos GP (1993) Meta-analysis of radiation therapy with and without adjuvant chemotherapy for malignant gliomas in adults. *Cancer* 71: 2585–2597

2   Holsi P, Sappino AP, de Tribolet N, Dietrich PY (1998) Malignant glioma: should chemotherapy be overthrown by experimental treatments? *Ann Oncol* 6: 589–600

3   Fortin D, Neuwelt EA (2002) *Therapeutic manipulation of the blood-brain barrier. Neurobase-neurosurgery.* 1st ed. Medlink CD-ROM, Medlink Corporation.

4   Zlokovic BV, Apuzzo ML (1997) Cellular and molecular neurosurgery: pathways from concept to reality—part I: target disorders and concept approaches to gene therapy of the central nervous system. *Neurosurgery* 40: 789–80

5   Kroll RA, Neuwelt EA (1998) Outwitting the blood-brain barrier for therapeutic purposes: Osmotic opening and other means. *Neurosurgery* 42: 1083–1100

6   Rapoport SI, Hori M, Klatzo I (1972) Testing of a hypothesis for osmotic opening of the blood-brain barrier. *Am J Physiol* 223: 323–331

7   EA Neuwelt (ed): *Implications of the Blood-Brain Barrier and Its Manipulation.* New York, Plenum Press, 1989, vol 1 and 2.

8   Pardridge WM (1991) Advances in cell biology of blood-brain barrier transport. *Semin Cell Biol* 2: 419–426

9   Pardridge WM (2002) Targeting neurotherapeutic agents through the blood-brain barrier. *Arch Neurol* 59: 35–40

10   Pardridge WM (1997) Drug delivery to the brain. *J Cereb Blood Flow Metab* 17: 713–731

11   Bobo RH, Laske DW, Akbasak A, Morrison PF, Dedrick RL, Oldfield EH (1994) Convection-enhanced delivery of macromolecules in the brain. *Proc Natl Acad Sci USA* 91: 2076–2080

12   Lieberman DM, Laske DW, Morrison PF, Bankiewicz KS, Oldfield EH (1995) Convection-enhanced distribution of large molecules in gray matter during interstitial drug infusion. *J Neurosurgery* 82: 1021–1029

13   Rainov NG, Kramm CM (2001) Vector delivery methods and targeting strategies for gene therapy of brain tumors. *Curr Gene Ther* 1: 367–383

14   Kroll RA, Pagel MA, Muldoon LL, Roman-Goldstein S, Neuwelt EA (1996) Increasing volume of distribution to the brain with interstitial infusion: dose, rather than convection, might be the most important factor. *Neurosurgery* 38: 746–752

15   Muldoon LL, Nilaver G, Kroll RA, Pagel MA, Breakefiled XO, Chioca EA, Davidson BL, Weissleder R, Neuwelt EA (1995) Comparison of intracerebral inoculation and osmotic blood-brain barrier disruption for delivery of adenovirus, herpesvirus, and iron oxide particles to normal rat brain. *Am J Pathol* 147: 1840–1851

16   Bartus RT, Snodgrass P, Dean RL, Kordower JH, Emerich DF (2000) Evidence that Cereport's ability to increase permeability of rat gliomas is dependent upon extent of tumor growth: implications for treating newly emerging tumor colonies. *Exp Neurol* 161: 234–44

17   Brem H, Ewend MG, Piantadosi S, Greenhoot J, Burger PC, Sisti M (1995) The safety of interstitial chemotherapy with BCNU-loaded polymer followed by radiation in the treatment of newly diagnosed malignant gliomas: Phase I trial. *J Neurooncol* 26: 111–123

18   Wang PP, Frazier J, Brem H (2002) Local drug delivery to the brain. *Adv Drug Deliv Rev* 54: 987–1013

19   Bradbury MWB (1986) Appraisal of the role of endothelial cells and glia in barrier breakdown. In: AJ Suckling, MG Rumsby, MWB Bradbury (eds): *The Blood-Brain Barrier in Health and Disease*. Ellis Horwood, Chichester, 128–129

20   Selman WR, Lust WD, Ratcheson RA (1996) Cerebral blood flow. In: RH Wilkins, SS Rengachary (eds): *Neurosurgery*. McGraw-Hill, New York, 1997–2007

21   Broman T, Olsson O (1949) Experimental study of contrast media for cerebral angiography with reference to possible injurious effects on the cerebral blood vessels. *Acta Radiol* 31: 321–334

22   Rapoport SI, Hori M, Klatxo I (1972) Testing of a hypothesis of osmotic opening of the blood-brain barrier. *Am J Physiol* 223: 323–331

23   Brightman MW, Hori M, Rapoport SI, Reese TS, Westergaard E (1973) Osmotic opening of tight junctions in cerebral endothelium. *J Comp Neurol* 152: 317–325

24   Dorovini-zis K, Bowman PB, Betz AL, Goldstein GW (1984) Hyperosmotic arabinose solutions open tigh junctions between brain capillary endothelial cells in tissue culture. *Brain Res* 302: 383–386

25   Rapaport SI, Fredericks WR, Ohno K, Pettigrew KD (1980) Quantitative aspects of reversible osmotic opening of the blood-brain barrier. *Am J Physiol* 238: R421–R431

26   Gummerlock MK, Neuwelt EA (1990) The effects of anesthesia on osmotic blood-brain barrier disruption. *Neurosurgery* 26: 268–277

27   Remsen LG, Pagel MA, McCormack CI, Fiamengo S, Sexton G, Neuwelt EA (1999) The influence of anesthetic choice, $PaCO_2$, and other factors on osmotic blood-brain barrier disruption in rats with brain tumor xenografts. *Anesth Analg* 88: 559–567

28   Fortin D, McCormick CI, Remsen LG, Nixon R, Neuwelt EA (2000) Unexpected neurotoxicity of etoposide phosphate administered in combination with other chemotherapeutic agents after blood-brain barrier modification to enhance delivery, using propofol for general anesthesia, in a rat model. *Neurosurgery* 47: 199–207

29   Bhattacharjee AK, Nagashima T, Kondoh T, Tamaki N (2001) Quantification of early blood-brain barrier disruption by *in situ* brain perfusion technique. *Brain Res Prot* 8: 126–131

30   Blasberg RG, Groothuis D, Molnar P (1990) A review of hyperosmotic blood-brain bar-

rier disruption in seven experimental brain tumor models. In: BB Johansson, C Owman, H Widner (eds): *Pathophysiology of the Blood-Brain Barrier.* Elsevier, Amsterdam, 197–220

31 Markowsky SJ, Zimmerman CL, Tholl D, Soria I, Castillo R (1991) Methotrexate disposition following disruption of the blood-brain barrier. *Ther Drug Monit* 13: 24–31

32 Robinson PJ, Rapoport SI (1991) Model for drug uptake by brain tumors: Effects of osmotic treatment and of diffusion in brain. *J Cereb Blood Flow Metab* 11: 165–168

33 Remsen LG, McCormick CI, Sexton G, Pearse HD, Garcia R, Neuwelt EA (1995) Decreased delivery and acute toxicity of cranial irradiation and chemotherapy given with osmotic blood-brain barrier disruption in a rodent model: The issue of sequence. *Clin Cancer Res* 1: 731–739

34 Kramer S (1968) The hazards of therapeutic irradiation of the central nervous system. *Clin Neurosurg* 15: 301–318

35 Doran SE, Ren XD, Betz AL, Pagel MA, Neuwelt EA, Roessler BJ, Davidson BL (1995) Gene expression from recombinant viral vectors in the CNS following blood-brain barrier disruption. *Neurosurgery* 36: 965–970

36 Muldoon LL, Nilaver G, Kroll RA, Pagel MA, Breakefield XO, Chiocca EA, Davidson BL, Weissleder R, Neuwelt EA (1995) Comparison of intracerebral inoculation and osmotic blood-brain barrier disruption for delivery of adenovirus, herpesvirus and iron oxide particles to normal rat brain. *Am J Pathol* 147: 1840–1851

37 Neuwelt EA, Barnett PA, Hellström I, Hellström KE, Beaumier P, McCormick CI, Weigel RM (1988) Delivery of melanoma-associated immunoglobulin monoclonal antibody and Fab fragments to normal brain utilizing osmotic blood-brain barrier disruption. *Cancer Res* 48: 4725–4729

38 Neuwelt EA, Specht HD, Barnett PA, Dahlborg SA, Miley A, Larson SM, Brown P, Eckerman KF, Hellström KE, Hellström I (1987) Increased delivery of tumor-specific monoclonal antibodies to brain after osmotic blood-brain barrier modification in patients with melanoma metastatic to the central nervous system. *Neurosurgery* 20: 885–895

39 Neuwelt EA, Weissleder R, Nilaver G, Kroll RA, Roman-Goldstein S, Szumowski J, Pagel MA, Jones RS, Remsen LG, McCormick CI et al (1994) Delivery of virus-sized iron oxide particles to rodent CNS neurons. *Neurosurgery* 34: 777–784

40 Nilaver G, Muldoon LL, Kroll RA, Pagel MA, Breakefield XO, Davidson BL, Neuwelt EA (1995) Delivery of herpesvirus and adenovirus to nude rat intracerebral tumors following osmotic blood-brain barrier disruption. *Proc Natl Acad Sci USA* 92: 9829–9833

41 Doolittle ND, Miner ME, Hall WA, Siegal T, Hanson EJ, Osztie E, McAllister LD, Bubalo JS, Kraemer DF, Fortin D et al (2000) Safety and efficacy of a multicenter study using intraarterial chemotherapy in conjunction with osmotic opening of the blood-brain barrier for the treatment of patients with malignant brain tumors. *Cancer* 88: 637–647

42 Yang W, Barth RF, Carpenter DE, Moeschberger ML, Goodman JH (1996) Enhanced delivery of boronophenylalanine by means of intracarotid injection and blood-brain barrier disruption for neutron capture therapy. *Neurosurgery* 38: 985–992

43 Fortin D McAllister LD, Nesbit G, Doolittle ND, Miner M, Hanson JE, Neuwelt EA (1999) Unusual cervical spine cord toxicity associated with intra-arterial carboplatin, intra-arterial or intravenous etoposide phosphate, and intravenous cyclophosphamide with osmotic blood-brain barrier disruption in the vertebral artery. *Am J Neuroradiol* 20: 1794–1802

44 Saris SC, Blasberg RG, Carson RE (1991) Intravascular streaming during carotid infusion:

demonstration in humans and reduction using diastole-phased pulsatile administration. *J Neurosurg* 74: 763–772

45   Roman-Goldstein S, Clunie DA, Stevens J, Hogan R, Monard J, Ramsey F, Neuwelt EA (1994) Osmotic blood-brain barrier disruption: CT and radionuclide imaging. *Am J Neuroradiol* 15: 581–590

46   Kraemer DF, Fortin D, Doolittle ND, Neuwelt EA (2001) Association of total dose intensity of chemotherapy in promary central nervous system lymphoma and survival. *Neurosurgery* 48:1033–1041

47   Siegal T, Rubinstein R, Bokstein F, Schwartz A, Lossos A, Shalom E, Chisin R, Gomori JM (2000) *In vivo* assessment of the window of barrier opening after osmotic blood-brain barrier disruption in humans. *J Neurosurg* 92: 599–605

48   Rapoport SI (1996) Modulation of blood-brain barrier permeability. *J Drug Target* 3: 417–435

49   McAllister LD, Doolittle ND, Guastadisegni PE, Kraemer DF, Lacy CA, Crossen JR, Neuwelt EA (2000) Cognitive outcomes and long-term follow-up results after enhanced chemotherapy delivery for primary central nervous system lymphoma. *Neurosurgery* 46: 51–61

50   Doolittle ND, Muldoon LL, Brummett RE, Tyson RM, Lacy C, Bubalo JS, Kraemer DF, Heinrich MC, Henry JA, Neuwelt EA (2001) Delayed sodium thiosulfate as an otoprotectant agent against carboplatin-induced hearing loss in patients with malignant brain tumors. *Clin Cancer Res* 7: 493–500

51   Doolittle ND, Tyson RM, Lacy C, Quipotla JL, Bubalo JS, Kraemer DF, DeLoughery T, Neuwelt EA (2001) Potential Role of delayed high-dose sodium thiosulfate as protectant against carboplatin-based thrombocytopenia in patients with malignant brain tumors. *Neuro-Oncology* 3: 357

Progress in Drug Research, Vol. 61
(L. Prokai and K. Prokai-Tatrai, Eds.)
©2003 Birkhäuser Verlag, Basel (Switzerland)

Modifying peptide properties by prodrug design for enhanced transport into the CNS

By Katalin Prokai-Tatrai[1] and Laszlo Prokai[2]

[1]Department of Pharmacology and Therapeutics
College of Medicine
University of Florida
Gainesville FL 32610, USA
<kprokai@ufl.edu>
and
[2]Department of Medicinal Chemistry
College of Pharmacy and
The McKnight Brain Institute
University of Florida
Gainesville FL 32610, USA
<lprokai@grove.ufl.edu>

Katalin Prokai-Tatrai

was born and educated in Hungary. She received her Ph.D. in organic chemistry from the University of Veszprém in 1986. Shortly thereafter, she emigrated to the United States and became involved with organic and medicinal chemistry at the University of Florida, where she is currently on the research faculty of the Department of Pharmacology and Therapeutics in the College of Medicine. She has co-authored numerous papers in the fields of her principal research interest that includes CNS-targeting of neurochemicals, peptide chemistry, prodrugs and antioxidant neuroprotection. She is the proud mother of David (born 1983) and Marcel (born 1997).

Laszlo Prokai

received his Ph.D. in radiochemistry (1983), and Habilitation (2000) from the University of Veszprém in Hungary. He pursued postdoctoral research in Medicinal Chemistry and then joined the faculty of the College of Pharmacy at the University of Florida in 1991, where he is now Professor of Medicinal Chemistry, member of the McKnight Brain Institute, Scientific Advisor of the Biotechnology Program, and Affiliate Professor of the Department of Chemistry. His research interests include drug design and discovery, peptide transport and delivery into the central nervous system, biochemistry and molecular pharmacology of neuropeptides, mechanisms of estrogen neuroprotection, mass spectrometry of macromolecules and proteomics. Prof. Prokai has authored a book, co-authored nine book chapters, has over 100 publications in scientific journals, and has been an invited speaker on several national and international meetings. He serves on the Editorial Advisory Boards of Current Medicinal Chemistry – Central Nervous System Agents, Current Drug Delivery *and* Medicinal Chemistry Reviews – Online.

Contents

Key words

CNS-targeting, enkephalins, passive transport, prodrug, prodrug-amenable peptide analogue, thyrotropin-releasing hormone.

Glossary of abbreviations

Ala, alanyl; ACh, acetylcholine; ANOVA, analysis of variance; Arg, arginyl; BBB, blood-brain barrier; CDS, chemical delivery system; CNS, central nervous system; CSF, cerebrospinal fluid; DADLE, [D-Ala^2,D-Leu^5]enkephalin; EC, enzyme commission; ED_{50}, effective dose for 50% of the maximum activity; Gln, glutaminyl; Glu, glutamyl; Gly, glycyl; His, histidyl; IC_{50}, concentration to reach 50% inhibition; IAM, immobilized artificial membrane; i.p., intraperitoneal; i.v., intravenous; k'$_{IAM}$, IAM chromatography capacity factor; Leu, leucyl; LHRH, luteinizing hormone-releasing hormone; logP, logarithm of n-octanol/water partition coefficient; MRP, multidrug resistance-associated protein; PAM, peptidyl α-amidating monooxygenase; PAP, pyroglutamyl aminopeptidase; pGlu, pyroglutamyl; Phe, phenylalanyl; Pro, prolyl; P-gp, P-glycoprotein; POP, prolyl oligopeptidase; SEM, standard error of the mean; Ser, seryl; $t_{1/2}$, half-life; Thr, threonyl; TRH, thyrotropin-releasing hormone; TSH, thyrotropin-stimulating hormone; Tyr, tyrosyl; Val, valyl; Xaa, any amino acid residue.

1   Introduction

Many central nervous system (CNS) diseases and disorders remain very difficult to treat because of the limiting effect of blood-brain barrier (BBB) on delivery of drugs into the CNS. A limited brain-uptake prevents numerous, otherwise promising agents to become pharmaceutically useful entities. Neuropeptides, their analogues and peptidomimetics are typical of these agents. Besides their poor BBB penetration, size, innate water-solubility and absence of specific transport systems to ferry them into the brain parenchyma, most peptides also have short biological half-lives because of rapid metabolism and clearance from the body [1].

Due to these shortcomings, specific delivery strategies discussed in this book have been developed to potentially overcome the hurdle to peptide pharmacotherapy. Enhancing brain-uptake and bioavailability of peptides may be achieved by reducing their size through the removal of amino acid residue(s) or subunits that are not essential for biological activity. Replacement of polar, ionizable amino acid residues with non-polar ones (e.g. Leu) at the sites that do not interfere with binding to the target receptor could also enhance lipid-solubility essential for the transport through the BBB. Other molecular manipulations for increasing lipid-solubility, which ideally also improve metabolic stability, include N-alkylation of one or more amide nitrogens in the backbone of the peptide [2] or of amino-acid side-chains (e.g, converting Tyr to *ortho*-alkyl- or *ortho,ortho*-dialkyl-Tyr) [3], halogenation [4, 5], various amide-bond surrogates [6] and the introduction of unnatural amino acids such as lipoamino acids [7]. On the opposite end, several studies have demonstrated that O-linked glycosylation of peptides (on Ser or Thr residues) can promote their penetration across the BBB *via* the glucose transporter system (GLUT-1) [8]. There is no increase in the affinity of such glycopeptides to lipid membranes compared to the parent compound (the attachment of the carbohydrate moiety actually reduces lipophilicity), because an active transport process is responsible for ferrying the peptide drug into the CNS. Although strategies of producing appropriate peptide analogues by an irreversible alteration of the structure of the parent peptide generally yield an improvement in BBB-transport properties, they exploit specific knowledge on structure–activity relationships and rely on a hypothetical "degree of freedom" related to ligand/substrate binding to the target CNS-receptor. The coverage of this approach [9] is out of scope in this chapter. Our focus is on an overview of the prodrug strategies that bioreversibly alter the structure of the target peptide (either the native form or its analogue) for improved bioavailability/CNS-uptake.

2   The prodrug concept

The prodrug design is perhaps the most versatile chemical manipulation strategy that attempts to solve or reduce certain identifiable shortcomings (e.g. insufficient solubility, stability, bad taste, limited tissue-uptake, etc.) of a parent drug having limited clinical usefulness. This technique relies on bioreversible chemical alteration of the target agents to produce their prodrugs having improved physicochemical characteristics compared to those of the parent drugs [10, 11]. The term of "prodrug" or "proagent" was first introduced by Albert [12] in the late 1950's to define pharmacologically inactive chemical derivatives that could be used to alter the physicochemical properties of drugs, in a transient manner, to increase their usefulness and/or to decrease associated toxicity. Other names, such as "latentiated drugs" [13] or "congeners" [14] have also been used but "prodrug" is the most commonly accepted nomenclature.

There is no strict definition for this term; in a broad sense prodrugs can be described as precursors of the parent drugs having no intrinsic activity that must undergo enzymatic and/or chemical/spontaneous process in a (preferably) predictable way to regenerate the active agent *in vivo* (Scheme 1). Simple prodrugs contain a covalent link between the drug and the strategically selected chemical/transport moiety. However, these precursors of the target drugs are frequently obtained by multiple chemical manipulations; thus, the term of prodrug will be used here to refer to any inactive, bioreversibly modified lead compound independently from the number and nature of chemical manipulations and pro-moieties, respectively. A classical example for a widely used prodrug is aspirin that quantitatively releases the active agent (salicylic acid) *in vivo* by esterases, and was designed to be less corrosive to the gastrointestinal mucosa then its parent drug.

Site-specific drug delivery by prodrug design may be the ultimate goal, where the drug's therapeutic effect is exerted only at the "site-of action" with no or minimal effect elsewhere in the body. Thus, the inactive prodrug should ideally release the active agent either non-enzymatically at the site of action or by the action of target-specific enzymes that are either characteris-

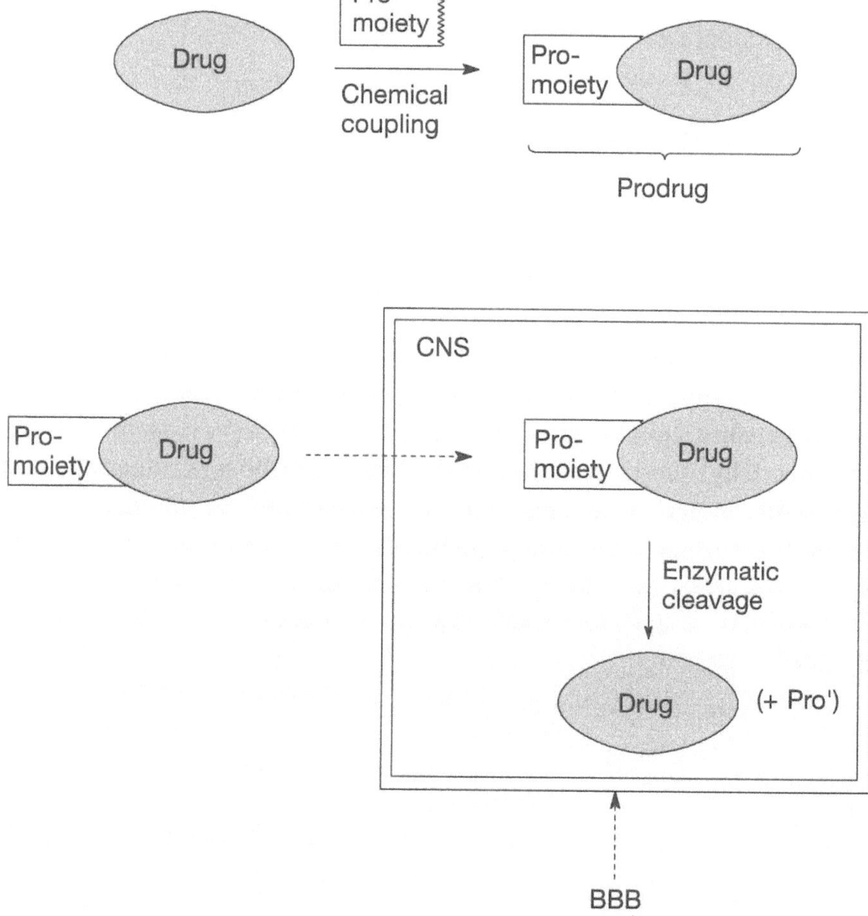

Scheme 1.
Schematic illustration of the prodrug concept.

tlc for or more abundant at the target site than anywhere else in the body. For true site-specific delivery or targeting a "lock-in" mechanism should prevent the efflux of the prodrug/drug from the target tissue [15] while the rapid elimination of the drug/prodrug from other body parts should be facilitated. Naturally, a prodrug must also have easy access to the target tissue. Numerous excellent and comprehensive reviews gave insight into prodrug design techniques [16–18].

3 Design of CNS-permeable prodrugs for peptides

The goal of CNS-targeting is to outwit the BBB. In general, the CNS-targeting prodrug strategy aims at a covalent and bioreversible attachment of lipophilic promoiety(ies) to the water-soluble peptide ("drug") to improve CNS-uptake and/or decrease proteolytic susceptibility in the plasma and interstitial fluids. Once the peptide prodrug crossed the BBB, "postbarrier" enzyme(s) or spontaneous processes regenerate the parent peptide. As it has clearly been stated in the previous chapters, the BBB is a very complex endothelial interface that regulates the exchange of drugs between CNS and the peripheral circulation acting as physical and enzymatic barriers. The flow is bi-directional, allowing the influx of certain molecules from the blood and the efflux of materials from the CNS. Most drugs exert their therapeutic effect in a concentration-dependent manner; thus, they need not only reach but also maintain a therapeutic concentration in the CNS.

In the classical sense, prodrugs are aimed at reaching the CNS by diffusion (passive transport), although promoieties that rely on active (carrier-mediated) transport have also been explored [18]. Passive transport through the BBB is controlled by several physicochemical parameters such as size (or rather, the molecular volume), charge and hydrogen-bonding capacity [19–21], yet lipophilicity (expressed as the logarithm of the n-octanol/water partition coefficient, logP) is generally considered the most important indicator for BBB-penetration [22, 23]. However, high lipid-solubility should also be avoided when making peptide prodrugs. A log P of 2 (i.e. 100-times higher affinity to the lipid-mimicking phase n-octanol than to water) is believed to be an optimal value for CNS-delivery [24]. Efflux mechanisms (e.g. a P-gp-mediated process [25]) operating in the BBB must be also considered in prodrug design, because they can eliminate the prodrug from the brain even in case of a robust influx [26, 27] resulting in poor CNS-retention and short biological half-life. In essence, the difference between the rate of influx across the BBB and that of the efflux (by, e.g., the cerebrospinal fluid (CSF)-sink, P-glycoprotein (P-gp) and multidrug resistance-associated protein (MRP)) of the prodrug, together with the *in situ* rate for the removal/conversion of the promoiety and the rate of the subsequent peptide degradation/elimination from its intended site of action, determines the concentration of the peptide in the CNS at any time after systemic administration.

Bioreversible lipidization of poorly CNS-available peptides involves chemical derivatization of the native peptide by taking advantage of the inherently present functional group(s) of the peptide chain. Those chemical "handles" may be the amino- and carboxy-termini or side chain amino-, hydroxyl-, thio- and carboxyl groups. Appropriate masking of these polar groups will also decrease hydrogen bonding capacity and render the prodrug neutral at physiological pH. Precise placement and choice of cleavable moieties can also provide protection against exo- and endopeptidases. In serum or plasma, many small peptides with free NH_2- and COOH-termini are degraded primarily by exopeptidases (amino- or carboxypeptidases) usually within a few minutes. Protection against protease recognition is one of the most important aspects of the CNS-targeting prodrug design for peptides, because lipid-soluble peptide prodrugs that can cross the BBB can only sustain adequate concentrations in the CNS if their blood concentration is maintained at sufficiently high levels by preventing their systemic degradation. As stated before, an increased lipophilicity through making a prodrug is expected to increase the brain uptake of an otherwise hydrophilic peptide. However, it may be not universally true. For example, an N-acetylated enkephalin analogue actually showed lower permeability across bovine endothelial-cell monolayers modeling the BBB *in vitro* than the parent peptide [28], which indicated that an increase in lipophilicity might be accompanied by increasing affinity to an efflux system in the CNS.

3.1 Synthesis of peptide prodrugs

Representative methods used for creating peptide prodrugs illustrated in Scheme 2 have previously been applied to small organic molecules possessing functional groups identical to those of peptides and, therefore, usually do not represent novel prodrug approaches. However, the presence of multiple functional groups could make peptide prodrug synthesis much more difficult.

Derivatization of the C-terminal- or side-chain carboxyl as well as of hydroxyl groups results in ester-type prodrugs (Scheme 2a) that have most frequently been shown as promising prodrug candidates because their synthesis is straightforward, they have reasonable chemical stability and readily hydrolyze *in vivo* due to the abundance of endogenous esterases in the body. The rate of either enzymatic or chemical hydrolysis may be controlled by appropriately chosen ester function. Simple esters can be obtained by condensation between an alcohol and an acid. This reaction is commonly carried out by using coupling agents, such as 1,3-dicyclohexylcarbodiimide (DCC) or benzotriazole-1-yl-oxy-tris-pyrrolidino-phosphonium hexafluorophosphate (PyBOP) in the presence of catalyst e.g. 4-(dimethylamino)pyridine (DMAP) either in solution phase or utilizing solid phase peptide synthesis (SPPS) in which the peptide/peptide prodrug is assembled on a polymer-support. Recently, we also used polymer-supported reagents (DCC and DMAP) efficiently instead of SPPS for the preparation of CNS-permeable [Glu^2]TRH prodrugs [29].

5-Oxazolidinone derivatization on the COOH-terminus (Scheme 2b) has also been suggested in peptide prodrug design [30, 31], primarily for protecting the peptide bond against enzymatic cleavage. The regeneration of the target peptide is based on the esterase-sensitivity of the lactone ring of oxazolidinone followed by the spontaneous decomposition of the N-hydroxyalkyl intermediate that possesses intrinsic chemical instability.

The design/synthesis of peptide prodrugs may be quite challenging sometimes. This is especially true, when masking of amino groups by acylation is attempted, because simple amides are mostly too stable *in vivo* (and chemically) to be useful prodrug forms [32]. Prodrug design relying on forming an amide bond between the peptide and the promoiety should aim at specific enzymes [33, 34] or specific chemistry [35, 36] (Scheme 2c). An amide-bond is formed upon reacting amines with acid, acid-halide or acid anhydride. With the latter two reagents, no additional coupling agent is necessary. Carbamate preparation on an amino group of the peptides has also been considered [37, 38], together with N-α-hydroxyalkyl and N-α-alkyloxycarbonyl derivatizations [39] (Scheme 2d). Preparing 4-imidazolidinones (Scheme 2e) may also offer a feasible solution for bioreversible amine-derivatization [40]. 4-Imidazolidinones are readily formed by condensing an α-aminoamide moiety such as N-terminal amine with carbonyl compounds (aldehydes or ketones). The regeneration of the parent peptide will depend on pH, the structure of the carbonyl compounds (R$_2$, R$_3$), and the parent peptide. Altogether, making prodrugs from peptides has been possible by a variety of ways (by methods illustrated in Scheme 2 individually or in various combinations to obtain, e.g. cyclic prodrugs) as described below in this chapter.

When the placement of the lipidizer/transport promoiety is intended within the peptide backbone (e.g. for protection against endopeptidases), either the side-chain functional groups (amino, hydroxyl, carboxyl, thiol or carboxyamide) or an auxiliary introduction of a functional group (e.g. by

a) Ester-type Prodrugs

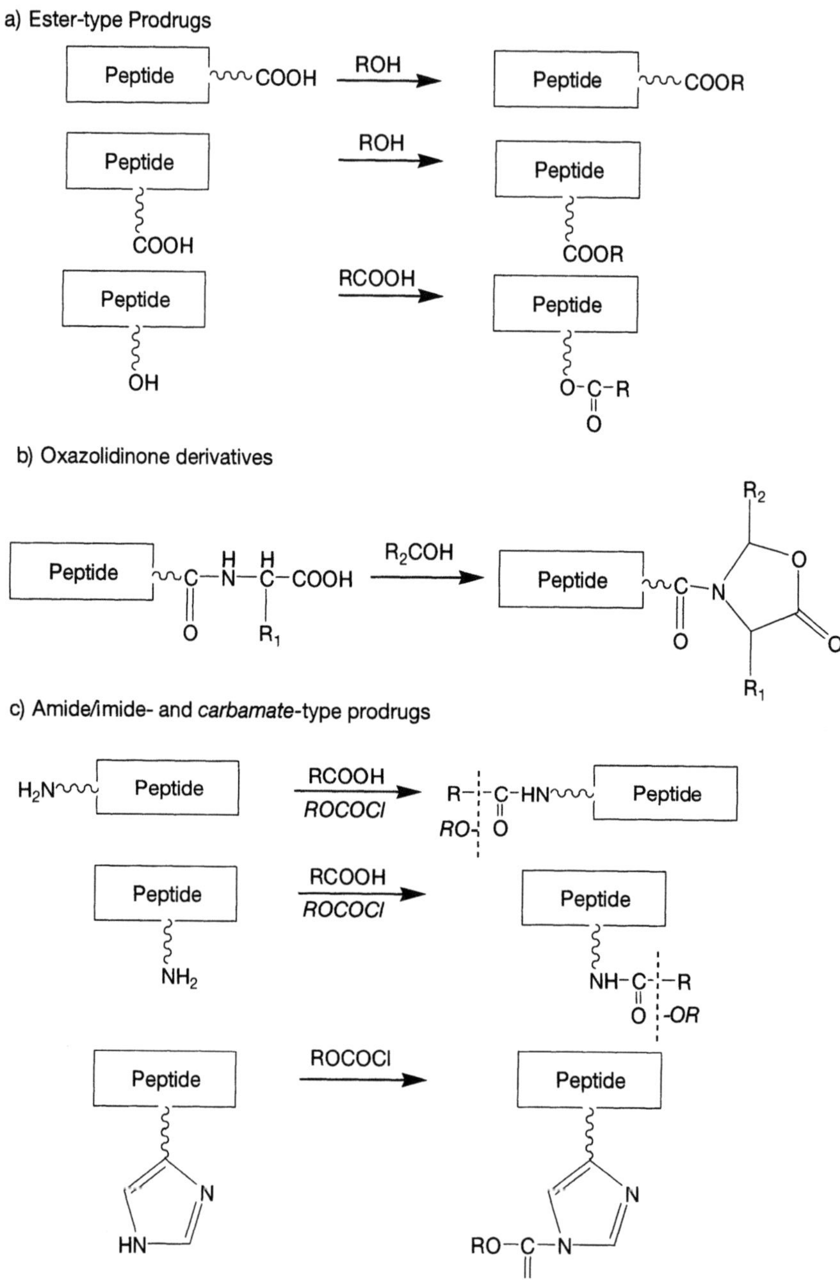

b) Oxazolidinone derivatives

c) Amide/imide- and *carbamate*-type prodrugs

Scheme 2.
Representative methods for peptide prodrug preparation.

d) N-α-hydroxyalkyl and N-α-acyloxyalkyl prodrug

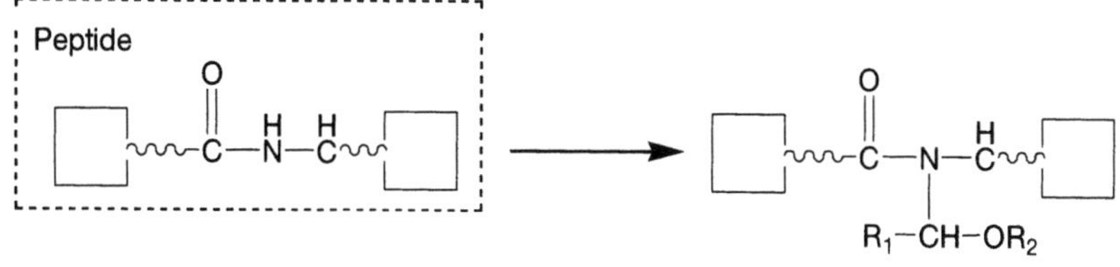

$R_2 = H$ or R_3-CO

e) 4-Imidazolidinone derivatives

Scheme 2 (continued).

hydroxymethylation of the nitrogen of the targeted peptide bond) may provide sites for chemical manipulation. Introduction of the promoiety may be done during the peptide-chain assembly involving continuous elongation or segment coupling or, alternatively, by direct derivatization of the parent peptide using solid- or solution-phase chemistry or the combination of both.

All the functional groups that are not to be involved in chemical reactions during peptide/prodrug synthesis should be suitably protected. The introduced promoiety of the prodrug should also be chemically resistant against the conditions through which the protecting groups necessary during the peptide synthesis are removed at the end of the synthetic procedure. These requirements could create great difficulties for the medicinal chemists. The feasibility to synthesize peptide prodrugs has often been a critical issue that limited progress beyond simple model peptides or even merely peptide models. Nevertheless, the application of the prodrug strategy has been promising for a variety of peptides, as discussed through representative examples in the following sections.

3.1.1 Peptide prodrugs by masking only a single functional group of the target peptide

Much attention has been paid to develop metabolically stable and CNS-available analgesics to avoid problems associated with the use of narcotics such as morphine. Enkephalins are endogeneous pentapeptides (Met/Leu-enkephalins, Tyr-Gly-Gly-Phe-Met/Leu) that bind to opioid receptors [41]. Their action on pain perception, addictive states and psychiatric disorders are well documented. However, analgesia is their best-known central effect [42] and its monitoring is the most commonly used pharmacological paradigm to follow opiate activity. The metabolic stability of the enkephalins is very poor, at least four different type of enzymes contribute to their rapid deactivation [43]. Therefore metabolically more stable synthetic analogues have been used as templates for CNS-permeable prodrug design.

One of those analogues is [D-Ala^2, Leu^5]enkephalin (DALE) to which adamantane-based moieties (e.g. 1-or 2-adamantyl) were attached *via* an ester-bound to the COOH-terminus [44, 45]. The bulky tricyclodecane cage structure (Fig. 1) gave an instant 100-fold increase, based on the logP, in lipid solubility to the prodrugs compared to DALE. CNS-delivery of DALE was evaluated by monitoring its antinociceptive activity (by the tail-pressure method in male mice) after subcutaneous administration of the prodrugs. The most potent prodrug in this assay had the 1-adamantyl tail on its C-terminus. This conjugate produced a significant antinociception, although a large dose was required (70 μmole/kg body weight). Nevertheless, the unmodified DALE was ineffective under the same experimental conditions. The maximum level of pain threshold was detected at 60–90 min post-administration, but no duration of action has been reported. While it is very conceivable that the measured analgesia is of central origin (i.e. naloxone-reversible) it has not been demonstrated experimentally. However, peptide conjugates were ineffective in the pharmacological paradigm used when the bulky adamantane-based group was placed on the NH$_2$-terminus (either as an amide or a carbamate) instead of the COOH-terminus. This may have been due to the excessive *in vivo* stability of the NH$_2$-terminal conjugate preventing the efficient removal of the promoiety necessary to allow for the subsequent binding to the opioid receptor.

Ester-type prodrugs have been recently evaluated for another Leu-enkephaline analogue Tyr-D-Ala-Gly-Phe-Leu-NH$_2$ (amidated at the COOH-ter-

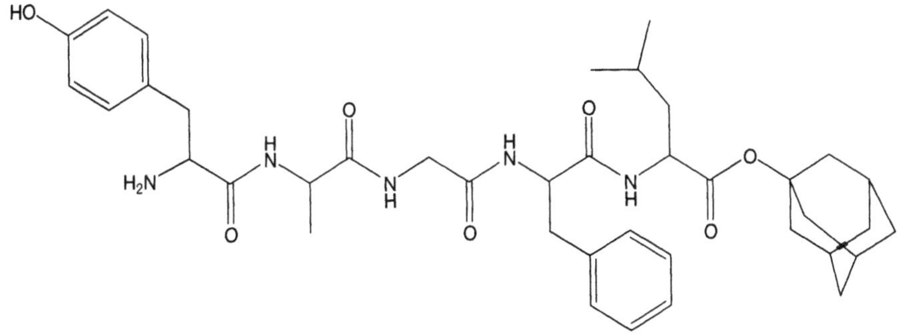

Figure 1.
1-Adamantyl ester prodrug of DALE.

minus for protection against carboxypeptidases) [46]. O-Acetyl, O-propionyl and O-pivaloyl esters (Fig. 2) were prepared on the tyrosine phenolic hydroxyl group of the N-protected peptide with the appropriate acid anhydride. *In vitro* studies revealed that the prodrugs degraded quantitatively in buffer and plasma, and as it was expected the most lipophilic and sterically hindered pivaloyl prodrug was the most stable in the biological media studied (e.g. $t_{1/2}$ of 2.6 h in human plasma). A 18-fold increase in the apparent permeability coefficient (P_{app}, basolateral to apical) was determined for this prodrug compared to the parent peptide in 30% Caco-2 cells at pH 7.4 and 37 °C. However, the parent peptide and its prodrugs were metabolized with similar half-lives in pure leucine aminopeptidase solution; thus, esterification of the phenolic hydroxyl group did not influence stability towards aminopeptidases. No pharmacological paradigm such as measurement of antinociception has been reported for assessing the *in vivo* usefulness of these prodrugs.

Thyrotropin-releasing hormone (TRH, pGlu-His-Pro-NH$_2$) and structurally related TRH-like peptides have been considered lead compounds for developing useful CNS agents [47]. The small tripeptide is the first hypothalamic releasing factor characterized, establishing the fundamental proof for the existence of a neuroendocrine regulation of pituitary functions by hypothalamic neuronal structures [48, 49]. A variety of behavioral effects are induced by peripheral and central application of TRH [47]. Therefore, it has been implicated in the management of various neurologic and neuropsychiatric disorders such as depression, brain injury, Alzheimer's disease and schizophrenia. The TRH-like tripeptide [Glu^2]TRH (pGlu-Glu-ProNH$_2$), although originally identified from rabbit prostate, has been shown to occur in the

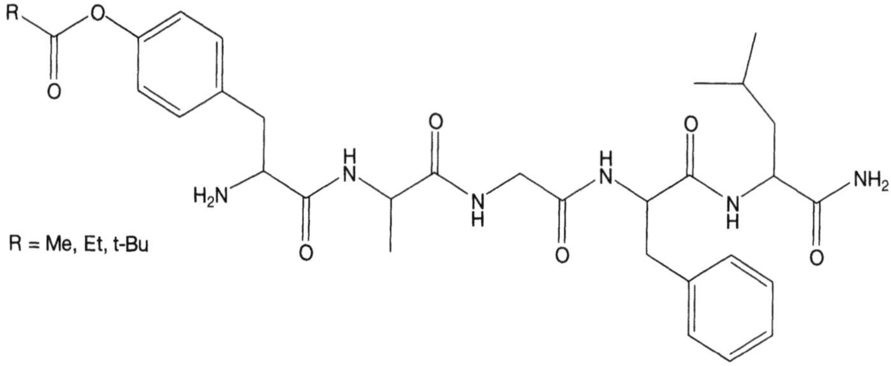

R = Me, Et, t-Bu

Figure 2.
Ester prodrugs of an enkephalin analogue, Tyr-D-Ala-Gly-Phe-Leu-NH$_2$.

human brain [50]. Although pharmacological activities are similar to those of TRH, the beneficial effects after treatment with [Glu^2]TRH are reportedly more robust or prolonged. The latter peptide is predominantly ionized at physiological pH at the carboxyl group in the side-chain of [Glu^2], which would prevent pharmacologically significant amount from entering the brain, by passive transport. Therefore, we synthesized several esters of [Glu^2]TRH (Fig. 3) and evaluated for their potential to interact with biological membranes by immobilized artificial membrane (IAM) chromatography, for their *in vitro* stability and for analeptic (CNS) activity, as summarized in Table 1 [29]. IAM chromatography measures the partitioning into monolayers of cell membrane phospholipids immobilized by covalent binding on silica particles. The chromatographic capacity factor (k'_{IAM}) for a compound obtained by the method is directly related to its partition coefficient between the aqueous phase and the chemically bonded membrane phase and, ultimately, to the K_m value representing its fluid membrane partition coefficient [51, 52]. All ester prodrugs showed, as expected, increased membrane affinity compared to the [Glu^2]TRH, and the hexyl ester (R = n-Hex) yielded the highest k'_{IAM} (16.0).

In vitro stability studies in mouse brain homogenate (20%, w/v) revealed that the half-lives were 20 and 22 min when R = Me and n-Hex, respectively, 25 min for R = c-Hex, and 70 min for the sterically hindered ester (R= t-Bu). On the other hand, benzyl ester (R = Bz) was quite stable in the tissue ($t_{1/2} > 2$ h) that rendered the compound practically useless as a prodrug.

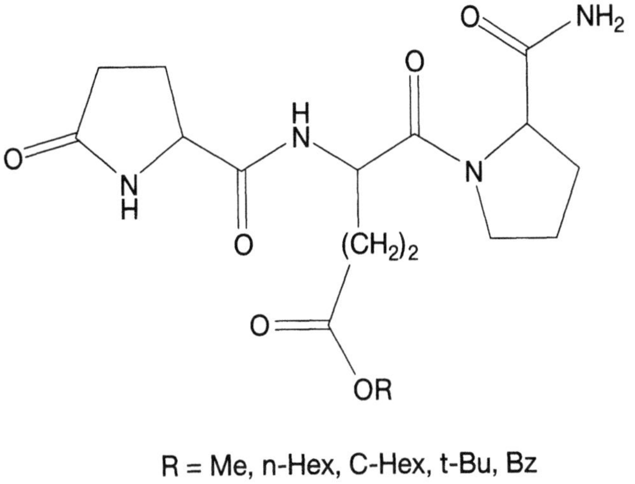

R = Me, n-Hex, C-Hex, t-Bu, Bz

Figure 3.
Various ester prodrugs of [Glu^2]TRH.

The antagonism of the barbiturate-induced narcosis in mice, a convenient pharmacological paradigm for the evaluation of TRH-related peptides, was used to assess the increase in the access of the prodrugs to the CNS at 10 µmole/kg body weight, i.v. When compared to an equimolar dose of the parent peptide, prodrugs having R = Me, n-Hex, c-Hex, respectively, showed a statistically significant decrease in the sleeping time. Based on measuring analeptic activity in the animal model selected, these prodrugs also outperformed the parent compound injected at 10-times higher dose (100 µmole/kg body weight). The measured pharmacological effect appeared to correlate well with the *in vitro* metabolic stability of the prodrugs in mouse brain homogenate. A slight influence of the increase in membrane affinity was also revealed upon comparing the analeptic response of methyl and hexyl esters (Tab. 1). Therefore, esterification of the COOH in the side chain of the Glu residue with primary alcohols (the more lipophilic the better) afforded the most promising prodrugs of [Glu^2]TRH. Dose-response studies with the hexyl ester also revealed that ED$_{50}$ was around 2 µmole/kg body weight–a ten-fold improvement compared to that of the parent peptide.

An interesting triglyceride-ester based prodrug for deltorphin II (Tyr-D-Ala-Phe-Glu-Val-Val-Gly, a δ-selective opioid agonist) with a molecular weight over 2500 Da was designed by Patel et al. for CNS-targeting [53]

Table 1.
Membrane affinity, *in vitro* half-life and analeptic effect (after i.v. administration into mice at equimolar doses of 10 μmole/kg body weight) of [Glu^2]TRH and its esters (Fig. 3) [29]

R	k'_{IAM}[a]	$t_{1/2}$ (min) in mouse brain homogenate (20%, w/w)	Sleeping time (± SEM) (min)[b,c]
H	0	–	65.3 ± 2.6*
Me	0.13	20	52.6 ± 1.3*
t-Bu	1.67	70	57.5 ± 1.3*
c-Hex	6.02	25	54.9 ± 2.3*
n-Hex	16.00	22	50.3 ± 2.1*
Bz	5.48	>120	72.0 ± 1.5

[a]Capacity factor measured by IAM chromatography.
[b]Pentobarbital (i.p., 60 mg/kg body weight) was administered 10 min after the injection of saline (vehicle), [Glu^2]TRH and its esters, respectively.
[c]Asterisks indicate statistically significant differences (ANOVA followed by Dunnett's test, $p < 0.05$) compared to the control group (sleeping time 79.7 ± 1.2 min).

(Fig. 4). It was postulated that the first step in the sequential metabolism leading to the release of deltorphin II was the cleavage of Arg-Pro by protease before or during absorption of the intact prodrug at the BBB, followed by hydrolysis of the lipid bond by lipases during or after transmission through the BBB. The prodrug produced a significant and prolonged (>4 h) analgesia in a hot plate assay at a dose of 60 μmole/kg body weight (i.p.) in mice, albeit the onset of action was slow. The conjugate (unlike the parent peptide) was not active in mouse vas deferens and guinea pig ileum bioassays and had very weak binding to the δ- and μ-opioid receptors; thus, it demonstrated the attributes of a genuine prodrug for deltorphin II. It was not clear from the report, however, that naloxone had been used to prove the central origin of the antinociception. The precise mechanisms of CNS-delivery and drug release from the prodrug, as well as the potential utility of this prodrug concept are yet to be further explored.

4-Imidazolidinone derivatives of Met- and Leu-enkephalins made by using various aldehydes and ketones as reagents during synthesis have also been studied as potential prodrugs [40, 54–56] to circumvent the rapid N-terminal metabolic inactivation by aminopeptidases. A large increase in membrane penetration characterized by the chromatographic capacity factors (k′) was achieved by the prodrugs (e.g., the 4-methyl cyclohexanone prodrug shown in Figure 5 had log k′ of 1.44) compared to Leu-enkephalin (log k′ = −0.15) [56]. Stability studies on the prodrugs indicated that all 4-imida-

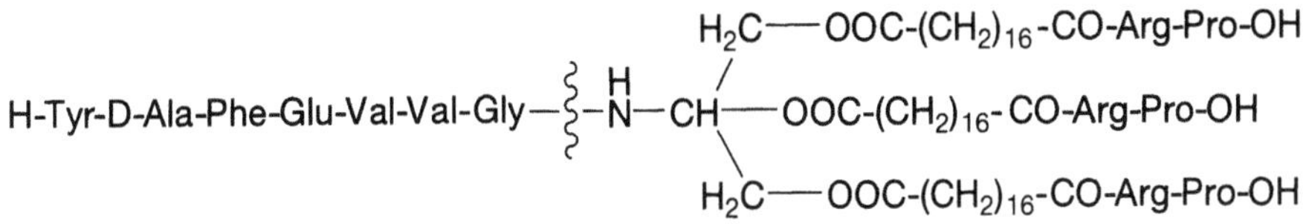

H-Tyr-D-Ala-Phe-Glu-Val-Val-Gly—N—CH with three branches:

$$\text{H}_2\text{C}\!-\!\!-\text{OOC-(CH}_2)_{16}\text{-CO-Arg-Pro-OH}$$
$$\text{H-Tyr-D-Ala-Phe-Glu-Val-Val-Gly}\!-\!\overset{\text{H}}{\text{N}}\!-\text{CH}\!-\!\!-\text{OOC-(CH}_2)_{16}\text{- CO-Arg-Pro-OH}$$
$$\text{H}_2\text{C}\!-\!\!-\text{OOC-(CH}_2)_{16}\text{-CO-Arg-Pro-OH}$$

Figure 4.
Triglyceride-ester based prodrug for deltorphin II.

zolidinone prodrugs degraded stoichiometrically to Leu-enkephalin in 0.02 M phosphate buffer at 37 °C, but were very stable ($t_{1/2} > 10$ h) against aminopeptidase N found in plasma and in the BBB, as well against angiotensin converting enzyme primary responsible for inactivating this endogenous peptide. Thus, based on lipid-solubility and metabolic stability, this type of prodrugs would offer a promising lead for CNS-targeting of enkephalins. However, *in vivo* performance has not been evaluated.

A classical example for a simple lipidization is the derivatization of the N-terminal pyroglutamine by N-acylation, N-acyloxymethylation for neuropeptides like TRH, luteinizing hormone-releasing hormone (LHRH), neurotensin, bombesin, and gastrin [57, 58]. These N-terminal modifications not only increase peptide lipophilicity, but also protect against pyroglutamyl aminopeptidase (PAPase I) partly responsible for the short half-live of these peptides. For example, TRH has a $t_{1/2}$ of only 6 to 8 min after parental administration in humans, mainly because of the TRH-specific serum enzyme PAPase II [47] that hydrolyzes the bond between pGlu and His. Therefore, disguising the proximity of this bond with lipidizers is expected not only to enhance brain-uptake of TRH, but also to protect against PAPase II.

TRH and its peptide models were extensively studied by Bundgaard and co-workers for the attachment of promoieties to pGlu [57, 58] and to imidazole ring of the central His [38, 59, 60]. Using L-pGlu benzylamide, a good substrate for PAPase I ($t_{1/2} = 10$ min), they investigated N-acyl-, N-acyloxymethyl- and N-aminomethyl (N-Mannich bases) derivatizations on pGlu as potential prodrugs forms. The stability of these compounds was assessed in buffer, human plasma and in the presence of PAPase I (from calf liver). While the derivatives of the TRH-model were totally resistant against this enzyme, they did convert to the parent compound either spontaneously (Mannich base) or by plasma enzymes (N-acyl derivatives). N-hydroxymethyl, N-acetoxymethyl- and N-phthalidyl derivatives of the peptide model

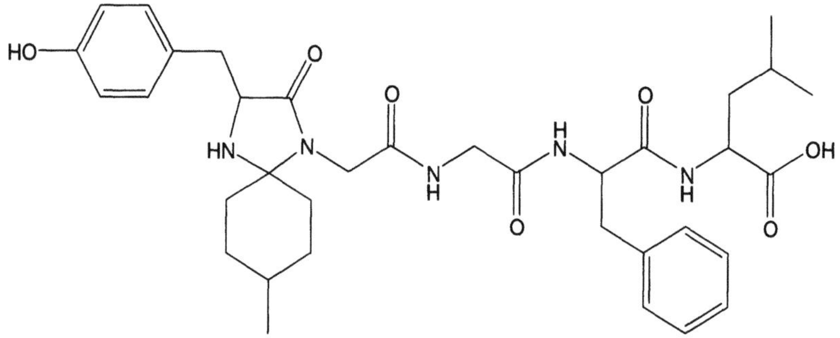

Figure 5.
A 4-imidazolidinone derivative of Leu-enkephalin.

were also found to be stable against the TRH-degrading enzyme and behaved as prodrugs in plasma, but the N-phenoxycarbonyl derivative manifested hydrolytic ring opening of the pyrrolidinone and did not regenerate L-pGlu benzylamide. Although these *in vitro* experiments on a peptide model have shown that certain chemical manipulations on pGlu are promising to obtain prodrugs to prevent proteolytic recognition, many studies would be necessary to validate their *in vivo* usefulness and chemical feasibility for actual peptides.

The imidazole group of the central His of TRH was also manipulated similarly to those of pGlu. Various chloroformates produced N-alkyloxycarbonyl prodrugs [38] (Fig. 6a) that resisted PAPase II, but quantitatively released TRH spontaneously or by esterase-catalyzed cleavage. They also possessed much higher lipid-solubility than TRH (e.g. the logP for octyloxycarbonyl derivative was 1.88, while the logP for TRH was –2.46). N-Phthalidylation as a promising approach for prodrug creation was also studied; however, like N-alkoxycarbonylation, it did not protect TRH against PAPase I or the intestinal prolyl endopeptidase [60].

In another study, lauroyl (Lau) group was introduced to pGlu of TRH [61, 62] (Fig. 6b). This highly lipid-soluble compound (with partition coefficient of 1.91 in n-octanol-buffer at pH 7.4, compared to 0.068 for TRH) was evaluated for CNS and endocrine activities. The relative potency of Lau-TRH (at a small dose of 0.25 μg/mouse, i.v.) on shortening the pentobarbital-induced sleeping time was decreased merely by 19%, while its TSH-releasing activity decreased by 36% compared to that of TRH. Lau-TRH also was more stable in rat plasma than TRH [62]. As far as TRH-related CNS-therapy is concerned, it

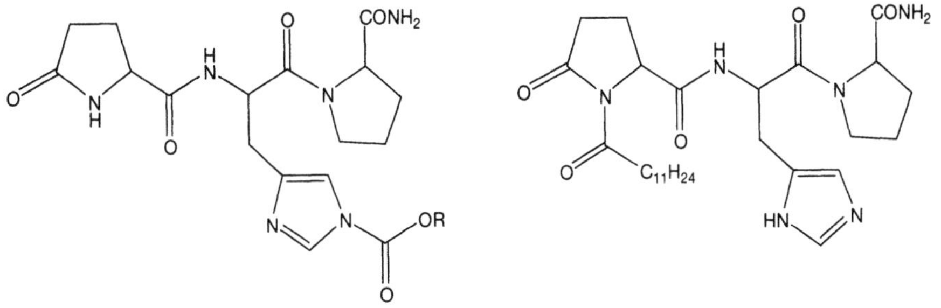

R = Me, Et, iPr, Bu, Pent, n-Hex, n-Oct, 2-Me-Hex, c-Hex, Bz

a   b

Figure 6.
TRH prodrugs. a) N-alkyloxycarbonyl derivatization on the imidazole ring of His, b) lauroyl group placed on the pGlu residue.

is very important to recognize that TRH is not only a neuropeptide but also an endocrine hormone. Therefore, several attempts have been made over the years to produce analogues or TRH-like peptides with limited or diminished endocrine activity and enhanced metabolic stability for potential CNS- application [47].

γ-Glutamyl transpeptidase has also been implicated as an enzyme for drug release from potential prodrugs [34]. Dermorphin was used as a target compound and its N-γ-glutamyl prodrug was prepared. Although the prodrug produced a naloxone-reversible antinociception (based on tail-flick test in rats), it required an almost 10-times higher dose to produce only half of the antinociception of the parent peptide.

NH_2- or COOH-terminal Phe-extension was used for the δ-opioid active [D-Pen^2, D-Pen^5]enkephalin (DPDPE) and [D-Pen^2, L-Cys^5]enkephalin (DPLCE) to study its utility for CNS-targeting [63]. The NH_2-terminal Phe extension only moderately changed the BBB permeability coefficient compared to DPDPE (to $57.94 \pm 2.78 \cdot 10^{-4}$ cm/min from $49.24 \pm 2.78 \cdot 10^{-4}$ cm/min), while receptor-binding studies revealed no significant affinity to opioid receptors for Phe^0-DPDPE. Inhibition of leucine aminopeptidase with bestatin in the serum increased the conversion time of Phe^0-DPDPE from 6.8 min to 92.2 min while inhibition of aminopeptidase M with amastatin in brain homogenate increased the conversion time of Phe^0-DPDPE from 3.9 min to over 450 min. Thus, the NH_2-terminal Phe^0-extended opioid peptides represented true DPDPE- and DPLCE-prodrugs activated by aminopeptidase(s). However, other Phe^0-opioid peptides did not produce prodrugs, because they manifested both binding affinities to the cognate receptor(s) and *in vitro* biological activities [64].

3.1.2 Peptide prodrugs by simultaneous masking of multiple functional groups of the target peptide

As the above examples have shown, prodrug design for the majority of peptides focuses only on the modification of a single functional group (e.g. N-terminal amino group) that may contribute to certain shortcomings (enzymatic susceptibility on the unmasked terminal end, insufficient lipophilicity, etc.) for efficient CNS-targeting. Cyclic prodrugs of linear peptides have also been used to alter physicochemical properties to allow for penetration into biological membranes. Heterodetic peptide prodrugs can be obtained through "bridges" (ester, ether, disulfide, thioether, etc.) between the NH_2- and COOH-termini (head-to tail cyclization gives homodetic peptide prodrugs). An elegant design has been recently introduced by developing esterase sensitive acyloxyalkoxy [65], phenylpropionic acid [66], coumarinic acid [67] and substituted coumarinic [68–70] linkers to obtain cyclic prodrugs for Leu-enkephalin and DADLE (Fig. 7). The bioconversion to the target peptide occurred in two steps; the hydrolysis of the ester bond between the C-terminal and the linker catalyzed by esterases was followed by the chemical degradation of the intermediate amide to enkephalin by, e.g. a "trimethyl-lock" facilitated lactonization [71, 72]. The cyclic prodrugs of these opioid peptides were shown to have favorable physicochemical properties (e.g. increased lipophilicity and no charge) for membrane permeation and unique solution structures (β-turns [63, 73]) that reduce their hydrogen bonding potential. They also had enhanced metabolic stability towards exo- and endopeptidases. *In vitro* stability studies revealed that some of the cyclic prodrugs (acyloxyalkoxy- and oxymethyl-modified coumarinic acid-based prodrugs, specifically) had a propensity to undergo a more rapid bioconversion in rat plasma and liver than in the brain. *In vitro* cell-culture experiments also indicated that the BBB permeation of the prodrugs might be significantly restricted due to their substrate activities for the P-gp active-efflux system [74]. As a consequence, only a very small amount of DADLE (less than the amount after the administration of the unmodified DADLE) was found in the

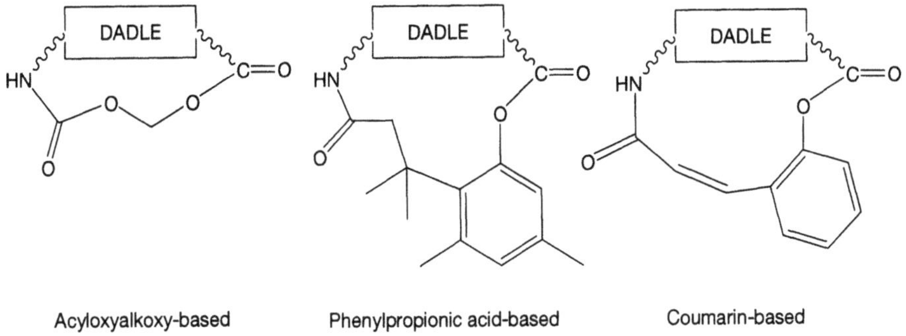

Figure 7.
Cyclic prodrugs of DADLE.

brain of experimental animals 10 min after the i.v. injection of the prodrug at a dose of 1 mg/kg body weight. Nevertheless, fine-tuning the linker pro-moieties may be a promising approach to be considered for the improvement of the biopharmaceutical properties once P-gp substrate properties are addressed upon designing the cyclic prodrugs.

A unique utilization of multiple chemical functionalization and specific enzymes such as NAD(P)H-dependent oxidoreductase [75], dipeptidyl dipep-tidases, proline oligopeptidase (POP) [76] or peptidyl glycine α-amidating monooxygenase (PAM) [77, 78] in the CNS-permeable prodrug design is the chemical delivery system approach (CDS) [33]. The CDS is a special prodrug that is capable of remarkable CNS-targeting, due to the "designed-in" reten-tion in the CNS, and has been applied to a wide variety of drugs [79]. The best-known element of the CDS strategy is a dihydropyridine moiety that enhances the lipid solubility of small molecules and is ubiquitously oxidized, analogously to the NAD(P)H coenzyme [75], to a membrane-impermeable, positively-charged pyridinium ion. Apparently, this oxidative process is responsible for the entrapment, after the CDS crosses the BBB, because the resultant pyridinium conjugates prevent drug/prodrug efflux by passive transport from the CNS. In the periphery the "oxidized" CDS is rapidly removed because pyridinium salts are easily eliminated from the body by the kidney or bile. Consequently, the concentration of the active drug is low in the periphery that, in turn, reduces systemic, dose-related toxicity and, thus, increases the therapeutic index. However, numerous issues (e.g. whether dihydropiridine-conjugates are substrates to any CNS-efflux system such as the P-gP) remain to be addressed by further studies.

For peptides, the simple CDS concept had to be modified, as shown in Figure 8. In this regard, the pivotal element of the design is a Pro- (or Ala-) based linker (sometimes called "spacer") capped with the 1,4-dihydrotrigonelline on the NH_2-terminus. Direct attachment of a dihydropyridine derivative to the NH_2-terminus of the peptide to be delivered into the CNS would not be suitable for prodrug design, because of the chemical/enzymatic stability of the amide bond. However, the enzymatic release of the core peptide in the CNS can be accomplished by appropriate peptidic (e.g. Pro/Ala-based) linkers utilizing specific enzymes. Lipidization on the COOH-terminus is also desirable to provide sufficient lipid-solubility to the entire conjugate for diffusion through the BBB.

After enzymatic oxidation of the dihydropyridine (to pyridinium) in the brain, a well-orchestrated sequential metabolism liberates the core peptide. The lipidizer on the COOH-terminal tail is hydrolyzed by esterases/lipases. The N-trigonellyl–linker moiety is then removed by dipeptidases (EC 3.4.14.2 and/or EC 3.4.14.5) if a single amino acid is used or by POP (EC 3.4.21.26) when a dipeptidyl moiety (Xaa-Pro/Ala) is the linker/spacer (S). These enzymes have specific cleaving sites (primarily after Pro, but also after Ala) in a peptide sequence [80] (The N-trigonellyl residue is believed to mimic a basic amino acid).

DALE and DADLE were the first peptides considered for the implementation of this strategy (Fig. 8) [33, 81]. The C-terminal esterification was carried out either with cholesterol (Cho, for DALE and DADLE) or 1-adamantaneethanol (Ada, for DADLE) in a customary manner (Scheme 2a). Because of the relative chemical instability of dihydropyridines (due to facile oxidation and pH-dependent degradation), synthetic procedures first aim at obtaining the corresponding pyridinium/trigonelline conjugate that can be easily reduced, e.g., with sodium dithionite [79] to the actual CDS. We also developed segment-coupling methods [81] that proved to be superior to sequential elongation for creating this type of pyridinium–peptide conjugates (Fig 8). *In vitro* stability studies conducted on the CDSs and their predicted biotransformation products in phosphate buffer, whole blood and 20% (w/w) brain homogenate showed that the CDSs were rapidly oxidized in tissue ($t_{1/2}$ around 2–3 min in blood and 35–40 min in brain homogenate, respectively) and the C-terminal lipidizer was removed within 15 min after administration. The sustained release of DADLE from the oxidized N-trigonellyl-linker-DADLE having free carboxyl end by dipeptidases was also unequivocally

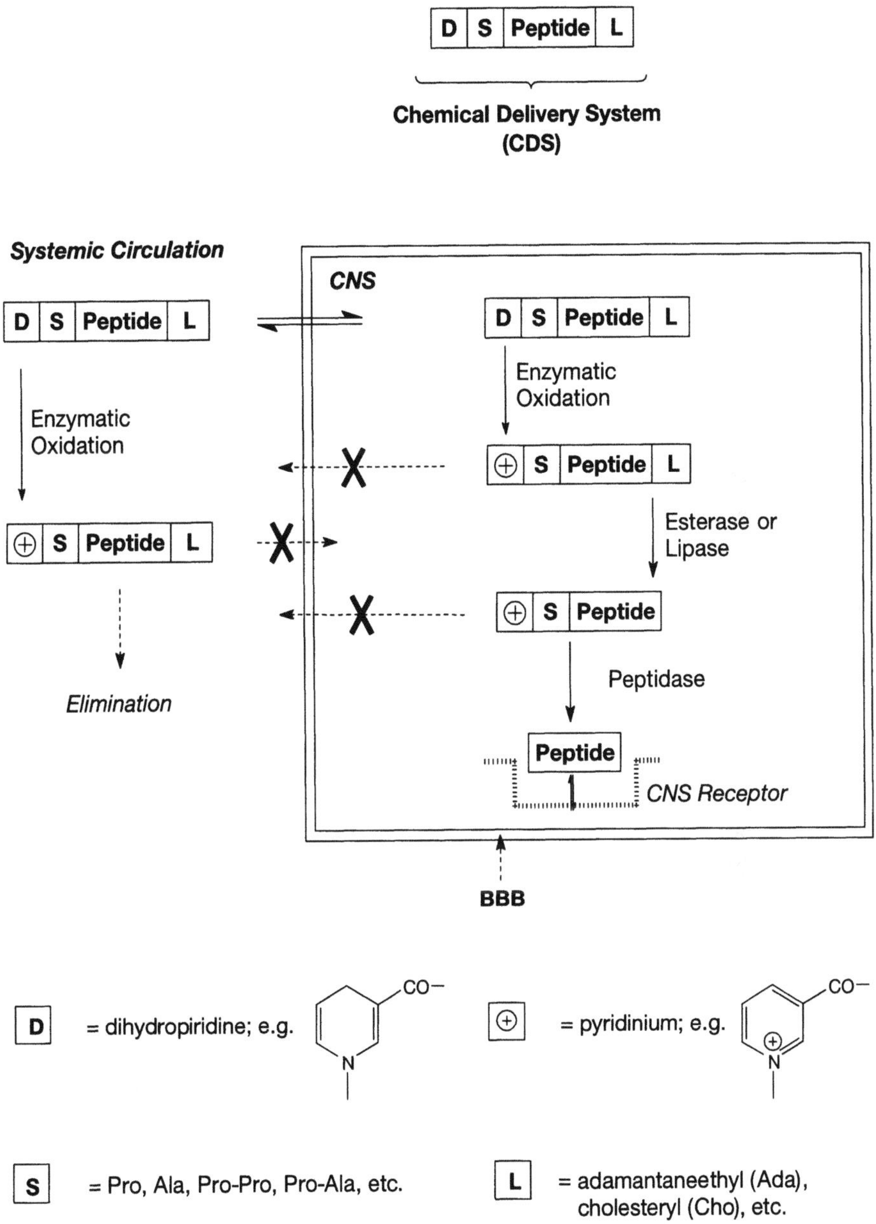

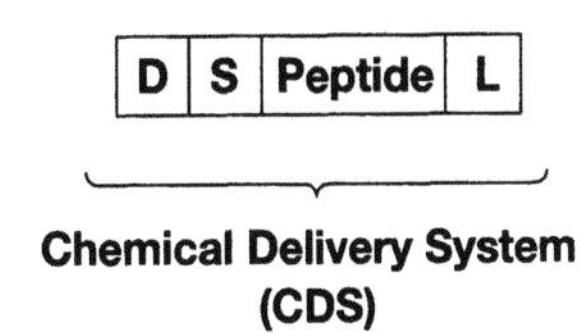

Figure 8.
Adaptation of the chemical delivery system concept to peptides.

proven by electrospray ionization mass spectrometric analysis. Receptor binding studies further confirmed that this novel concept for CNS-targeting of peptides, indeed, showed characteristics of true prodrugs (opiod-receptor binding affinity expressed as IC_{50} was 6 μM for the intact oxidized prodrug, 200 nM after the removal of Cho and 36 nM for DADLE, respectively). The efficacy of CDSs to deliver the peptides to the CNS was assessed ultimately by measuring CNS-mediated analgesia (using the tail-flick paradigm) upon i.v. administration to rats.

Table 2 shows that essentially no anticiception was observed with DALE embedded into the CDS–probably due to a fast metabolism ($t_{1/2}$ around 30 min *in vitro*) of the oxidized prodrug (N-Trigonellyl-Ala/Pro-DALE) to small non-opioid fragments. Thus, the bulky Cho group did not "disguise" the peptide against proteolytic cleavages, it had only a "lipidizer" function. However, a significant and long-lasting analgesia was produced by DADLE delivered to the brain by this CNS-targeting method (Tab. 2). The analgesia was naloxone-reversible but methylnaloxonium-irreversible confirming that central opioid receptors were solely responsible for the mediation of this pharmacological response. The use of the bulkier Cho group and Pro spacer in the molecules showed a better efficacy than those of Ada and Ala.

The peptide-CDS concept can be easily adopted for other peptides having amino- and carboxy-termini or side-chain functional groups (amino/imino and carboxyl/hydroxyl) that can provide synthetic handles for the functionalizations. However, peptides without these groups would seemingly defy a simple adaptation of the approach. A good example is TRH analogues in which the central His is replaced with a hydrophobic amino acid residue. [Leu^2]TRH has been a model peptide for our efforts to develop CNS-targeting CDSs, because it exerts a 2.5-fold increase in CNS activity and shows decreased TSH-releasing properties compared to TRH [82]. Essentially, we recognized that posttranslational cleavage of a large precursor polyprotein having 123 amino acid residues to produce TRH [83] involved PAM converting (in a two-step process proceeding through peptidyl-α-hydroxy-glycine) the COOH-terminal glycine residue of the Gln-His-Pro-Gly progenitor sequence to prolineamide. (To the analogy of this enzyme-catalyzed process, Bundgaard and co-workers employed hydroxyglycine extension of the COOH-terminus as a method to protect against α-chymotrypsin by using model peptides [84, 85]). Glutaminyl cyclase is responsile for the conversion of the NH$_2$-terminal Gln to pGlu [86, 87]. Therefore, we envisioned that Gln-Leu-Pro-Gly-OH would

Table 2.
Antinociception in male Sprague-Dawley rats after intravenous injection (4.2 μmole/kg body weight) of CDS (Fig. 8) for Leu-enkephalin analogues with varying COOH-terminal lipophilic ester (L) and dihydrotrigonellyl-to-peptide linker residue or "spacer" (S) functions [33, 81].

Enkephalin analogue	Lipophile (L)	Spacer (S)	% Anti-nociception (mean ± SD)	Time to reach maximum effect (min)	Duration of action (min)
DALE	Cho	Ala	minimal	n/a	transient
DADLE	Ada	Ala	31 ± 23	60	60–90
DADLE	Ada	Pro	53 ± 29	15	< 120
DADLE	Cho	Ala	46 ± 27	60	180–240
DADLE	Cho	Pro	51 ± 28	15–30	> 300

be, after its targeting to brain, a suitable progenitor peptide for [Leu^2]TRH. Since this precursor peptide possesses both NH_2- and COOH-termini for convenient chemical manipulations, we synthesized CDSs by the analogy of those of DADLE (Fig. 8). The N-1,4-dihydrotrigonelline was attached through a single amino-acid or dipeptide linker to NH_2-terminal of the Gln-Leu-Pro-Gly progenitor sequence and Gly was esterified with Cho (Fig. 9) [88]. The variation in the spacer moiety (considering Pro and Ala as building blocks) was intended to reveal the influence of peptide release on the efficacy [89].

Because of the complexity of the molecular architecture and the process of bioactivations shown in Figure 9, we thoroughly investigated the cascade of "designed-in" enzymatic reactions responsible for the release of the core peptide by using the appropriate synthetic peptide conjugates as substrates. One of the critical steps is the amidation of the oxidized CDS in the brain by PAM. *In vitro* studies on trigonellyl-Ala-Gln-Leu-Pro-Gly showed that this process (with $t_{1/2}$ 25–40 min *in vitro*) was significant only in the brain. In blood and liver the primary degradation occurred by the removal of Gly resulting in a Pro-OH (non-amidated) COOH-terminus that could only provide an inactive metabolite of the target peptide. The metabolism of trigonellyl-Ala-Gln-Leu-Pro-Gly was also slower in the liver than in the brain; thus, systemic formation of [Leu^2]TRH was practically prevented. The peptidolytic cleavage to form Gln-Leu-Pro-NH_2 and subsequent rapid formation of [Leu^2]TRH without significant side-reactions was also unequivocally detected upon incubation key intermediates (e.g. trigonellyl-Pro-Pro-Gln-Leu-Pro-NH_2) both by using a recombinant enzyme (POP) and in rat brain homogenate. Finally, an *in vivo* evaluation (analeptic activity in mice) pointed out

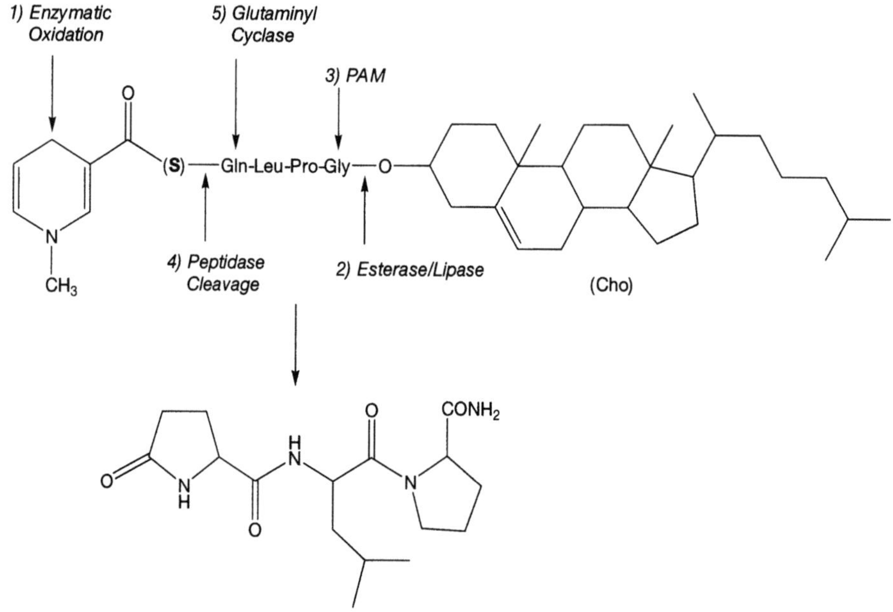

Figure 9.
Sequential metabolic regeneration of TRH from its CNS-permeable prodrug.

the critical influence of the peptidase cleavage affecting the Ala/Pro-Gln bond on the observed pharmacological effect, in which CDSs with Pro in the spacer (also preferred by the enzymes DP II/IV and POP) outperformed those with Ala-containing moieties. This observation has highlighted the need for a careful design and/or combinatorial optimization in linking the promoiety to the peptide based on considering the substrate properties of the enzyme(s) targeted for the cleavage *in vivo*.

Although it has not yet been studied for peptide CNS–targeting, an alternative carrier to the dihydropyridine → pyridinium system to produce retention in the CNS is noteworthy. Thiazolium precursors were applied for the brain-delivery of L-3,4-dihydroxyphenylalanine (DOPA), the precursor for the neurotransmitter dopamine [90]. The hypothesis behind this design is that disulfide containing moieties, such as thiamine disulfide attached *via* an ester bond to the carboxyl group of DOPA is bioreducible in the brain (opposite reaction to the dihydropyridine → pyridinium system that represents oxidation) faster than the hydrolyzes of the ester bond, thus, drug release.

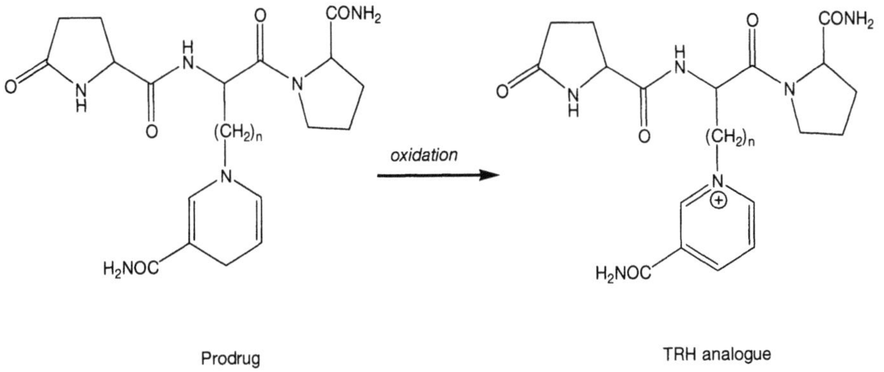

Figure 10.
Prodrug-amenable peptide analogues of TRH.

Among the prodrugs studied, 4-methylformylamino-3-(2-propyl)dithio-3-pentenyl-2-amino-3(3,4-dipivaloyloxyphenyl)propionate was found to deliver DOPA most efficiently to the brain at a dose of 25.4 μmole/kg body weight (i.v.) in rats with a peak concentration 15 min postadministration and producing 30- and 3.7-fold increase in the area-under-the-curve and mean residence time of DOPA, respectively, in the brain than DOPA itself.

3.2 Prodrug-amenable peptide analogues

Efforts to apply the prodrug strategy to CNS-targeting of peptides have also influenced the method to design CNS-active peptide analogues. Specific, pro-drug-amenable peptide analogues may be designed to facilitate transport across the BBB and/or retention in the CNS. For example, we have recently exploited the benefits of the CDS strategy in terms of distribution and pharmacokinetics/pharmacodynamics [91] by designing metabolically stable and exclusively CNS-active TRH analogues with the replacement of the peptide's [His2] residue with specific amino acid mimetics containing substituted pyridinium side-chains [92, 93] (Fig. 10). The most potent analogue in the series (n = 1) was shown to be TRH-equivalent but of a significantly longer duration of action than the parent peptide in antagonizing pentobarbital-induced narcosis, when administered i.v. in its prodrug form (dihydropyridine) into mice at a dose equimolar to that of TRH (Tab. 3). In rat plasma and brain homogenate (20%, w/v), the prodrugs converted to the pyridinium analogues with

Table 3.

Pharmacological comparison of the duration of analeptic action at equimolar dose (15 µmole/kg body weight, i.v.) of TRH and its pyridinium analogue administered as a dihydropiridine prodrug upon varying the time (Δt) for pentobarbital post-administration (i.p., 60 mg/kg body weight) [92, 93].

Δt (min)	Sleeping time after TRH (min, average $\pm$ SEM)[a]	Sleeping time after pyridinium analogue[b] (min, average $\pm$ SEM)[a]
10	24.1 $\pm$ 3.3	19.4 $\pm$ 2.1
20	37.5 $\pm$ 5.4*	13.9 $\pm$ 1.7
30	27.0 $\pm$ 4.0	14.2 $\pm$ 2.3
60	47.4 $\pm$ 5.3*	20.8 $\pm$ 3.8

[a]Asterisks indicate statistically significant differences (ANOVA followed by Dunnett's test, $p < 0.05$) compared to the experiment when cholinergic challenge was made 10 min after the injection of the prodrug for the TRH analogue.
[b]Figure 10, n = 1

half-lives around 20 min and 6 min, respectively. On the other hand, we determined $t_{1/2}$ of 16 min in brain homogenate and 11 min in plasma for TRH, while the analogues were very stable in these tissues (less than 10% degradation in 2 h). The longer $t_{1/2}$ of the prodrugs in plasma compared to $t_{1/2}$ in brain was also beneficial to the CNS-sequestration of the analogues after systemic administration. The maximal change in acetylcholine (ACh) concentration upon perfusion of the pyridinium-containing tripeptides into the hippocampus of rats was also achieved with analogue having the shortest side-chain (n = 1). No binding to the endocrine TRH-receptor was measured for these compounds; thus, the design afforded a novel lead for centrally acting TRH analogues. Besides TRH, a kyotorphin analogue with pyridinium side-chains and its prodrug have also been developed and evaluated [37].

4   Conclusions

The main goal of medicinal chemistry involving CNS-active peptides is to transform a lead molecule into a pharmaceutically useful drug candidate. This usually prompts chemical modifications, including the consideration to design prodrugs. Prodrugs are inactive chemical derivatives of the target peptide drugs that must undergo *in vivo* activation(s) for reverting to the active agents.

Peptide prodrug design usually aims at higher lipophilicity and improved metabolic stability enabling a better penetration and transport across the

BBB; however, the outcome of the prodrug approach is often unpredictable. Several criteria have to be fulfilled to create a useful, CNS-permeable prodrug for peptides. One of the most important features of the prodrug is the ability to be converted quantitatively to the parent drug *in vivo*. Simultaneously, the prodrug should not be substrate for any active transports present in the CNS that could prevent an otherwise robust accumulation of the prodrug/drug in the brain. Unfortunately, the increased lipophilicity of prodrugs may not only enhance CNS-uptake, but also could burden non-target tissue. It is also desirable that prodrugs have reasonable half-lives in plasma and not to be rapidly metabolized or cleared by the periphery, and protein-bound. A critical issue in the design of prodrugs for CNS-active peptides is, therefore, the choice and precise placement of the cleavable moiety or moieties to obtain efficacious absorption, distribution, metabolism and an optimal pharmacokinetic/pharmacodynamic profile. It is also clear that no unified method is possible for true CNS-targeting of all CNS-active peptides with potential clinical usefulness due to the complex structure of peptides/proteins and the intricate interplay among factors governing the entry and retention of these biomolecules in the CNS. Despite of advances in peptide pharmaceuticals, CNS-targeting of bioactive, stable and bioavailable peptide drugs has remained a very challenging task. Nevertheless, progress in this area has been steady and the continued exploration and development of the prodrug strategy outlined in this chapter are clearly warranted.

Acknowledgements

Part of the research presented here has been supported by grants (AG10485 and MH59360) from the National Institutes of Health (Bethesda, MD, USA).

References

1   Prokai L (1998) Peptide drug delivery into the central nervous system. In: EM Jucker (ed): *Progress in Drug Research*, Vol. 51, Birkhäuser, Basel, 96–131

2   Kawai M, Fukuta N, Ito N, Kagami T, Butsugan Y, Maruyama M, Kudo Y (1990) Preparation and opioid activities of N-methylated analogs of [D-Ala^2, Leu^5]enkephalin. *Int J Pept Prot Res* 35: 452–459

3   Witt KA, Slate CA, Egleton RD, Huber JD, Yamamura HI, Hruby VJ, Davis TP (2000) Assessment of stereoselectivity of trimethylphenylalanine analogues of δ-opioid [D-Phen(2), D-Pen(5)]-enkephalin. *J Neurochem* 75: 424–435

4   Weber SJ, Greene DL, Sharma SD, Yamamura HI, Kramer TH, Burks TF, Hruby VJ, Hersh LB, Davis TP (1991) Distribution, and analgesia of [H-3][D-Pen2, D-Pen5]enkephalin, and two halogenated analogs after intravenous administration. *J Pharm Exp Ther* 259: 1109–1117

5   Gentry CL, Egleton RD, Gillespie T, Abbruscato TJ, Bechowski HB, Hruby VJ, Davis TP (1999) The effect of halogenation on blood-brain barrier permeability of a novel peptide drug peptides. *Peptides* 20: 1229–1238

6   Delaet NGJ, Verheyden P, Velkeniers B, Hooghepeters EL, Bruns C, Tourwe D, Vanbinst G (1993) Synthesis, biological-activity and conformational study of a somatostatin hexapeptide analog containing a reduced peptide-bond. *Peptide Res* 6: 24–30

7   Wong A, Toth I (2001) Lipid, sugar and liposaccharide based delivery systems. *Curr Med Chem* 8: 1123–1136.

8   Polt R, Palian MM (2001) Glycopeptide analgesics. *Drugs Future* 26: 561–576.

9   Egleton RD, Davis TP (1997) Bioavailability and transport of peptides and peptide drugs into the brain. *Peptides* 18: 1431–1439

10   Bundgaard H (1992) Means to enhance penetration: 1. Prodrugs as a mean to improve the delivery of peptide drugs. *Adv Drug Del Rev* 8: 1–38

11   Stella V (1975) Pro-drugs: an overview and definition. In: T Higuchi, VJ Stella (eds): *Prodrugs as novel drug delivery systems.* ACS Symposium Series, Am Chem Soc, Washington DC ,1–115

12   Albert A (1958) A chemical aspect of selective toxicity. *Nature* 182: 421–423

13   Harper NJ (1959) Drug latentiation. *J Med Pharm Chem* 1: 467–500

14   Sinkula AA, Yalkowsky SH (1975) Rationale for design of biologically reversible drug derivatives: Prodrugs. *J Pharm Sci* 64: 181–210

15   Stella VJ, Himmelstein KJ (1980) Prodrugs and site-specific drug delivery. *J Med Chem* 23: 1276–1282

16   Stella VJ, Charman WNA, Naringrekar VH (1985) Prodrugs. Do they have advantages in clinical practice? *Drugs* 29: 455–473

17   Kearney AS (1996) Prodrugs and targeted drug delivery. *Adv Drug Deliv Rev* 19: 225–239

18   Anand BS, Dey S, Mitra AK (2002) Current prodrug strategies *via* membrane transporters/receptors. *Expert Opin Ther Biol* 2: 607–620

19   Levin VA (1980) Relationship of octanol/water partition coefficient, and molecular weight to rat brain capillary permeability. *J Med Chem* 23: 682–684

20   Pardridge WM (1999) Blood-brain barrier biology, and methodology. *J Neurovirol* 5: 556–569

21   Abraham MH, Chadha HS, Mitchell RC (1995) Hydrogen-bonding. Part 36. Determination of blood brain distribution using octanol-water partition coefficients. *Drug Dis Discov* 13: 123–131

22   Habgood MD, Begley DJ, Abbott NJ (2000) Determination of passive drug entry into the central nervous system. *Cell Mol Neurobiol* 20: 231–253

23   Banks WA, Kastin AJ (1985) Peptides and the blood-brain barrier: lipophilicity as a predictor of permeability. *Brain Res Bull* 15: 287–292

24   Hansch C, Björkroth JP, Leo A (1987) Hydrophobicity and central nervous system agents: on the principle of minimal hydrophobicity in drug design. *J Pharm Sci* 76: 663–687

25   Tatsuta T, Naito M, Oh-Hara T, Sugawara I, Tsuruo T (1992) Functional involvement of P-glycoprotein in blood-brain barrier. *J Biol Chem* 28: 20383–20391

26   Bak A, Gudmundsson OS, Friis GJ, Shiahaan TJ, Borchardt RT (1999) Acyloxy-alkoxy-based

cyclic prodrugs of opioid peptides: evaluation of the chemical and enzymatic stability as well as their transport properties across Caco-2 cell monolayers. *Pharm Res* 16: 24–29

27 Begley D (1996) The blood-brain barrier: principles for targeting peptides and drugs to the central nervous system. *J Pharm Pharmacol* 48: 136–146

28 Weber SJ, Abbruscato TJ, Browson EA, Lipkowski AW, Polt R, Misicka A, Haaseth RC, Bartosz H, Hruby VJ, Davis TP (1993) Assessment of an *in-vitro* blood-brain-barrier model using several [MET5] enkephalin opioid analogs. *J Pharm Exp* 266: 1649–1655

29 Prokai-Tatrai K, Nguyen V, Zharikova A D, Braddy A C, Stevens SM, Jr, Prokai L (2003) Prodrugs to enhance central nervous system effects of the TRH-like peptide pGlu-Glu-Pro-NH$_2$. B*ioorg Med Chem Lett* 13: 1011–1014

30 Buur A, Bundgaard, H (1988) Prodrugs of peptides. III. 5-oxazolidinones as bioreversible derivatives for the α-amido carboxy moiety in peptides. *Int J Pharm* 46: 159–167

31 Friis GJ, Bak A, Larsen BD, Frokjaer S (1996) Prodrugs of peptides obtained by derivatization of the C-terminal peptide bond in order to effect protection against degradation by carboxypeptidases. *Int J Pharm* 136: 61–69

32 Pitman IH (1981) Pro-drugs of amides, imides and amines. Med Res Rev 1: 189–214

33 Bodor N, Prokai L, Wu W-M, Farag H, Jonnalagadda S, Kawamura M, Simpkins J (1992) A strategy for delivering peptides into the central nervous system by sequential metabolism. *Science* 257: 1698–1700

34 Misicka A, Maszczynska I, Lipkowski AW, Stropova D, Yamamura HI, Hruby VJ (1996) Synthesis and biological properties of gamma-glutamyl-dermorphin, a prodrug. *Life Sci* 58: 905–911

35 Amsberry KL, Borchardt, RT (1990) The lactonization of 2'-hydrocinnamic acid amides: A potential prodrug for amines. *J Org Chem* 55: 5867–5877

36 Amsberry, KL, Gerstenberger, AE, Borchardt, RT (1991) Amine prodrugs which utilize hydroxy amide lactonization. II. A potential esterase-sensitive amide prodrug. *Pharm Res* 8: 455–461

37 Chen P, Bodor N, Wu WM, Prokai L (1998) Strategies to target kyotorphin analogs to the brain. *J Med Chem* 41: 3773–3781

38 Bundgaard H, Moss J (1990) Prodrugs of peptides. VI. Bioreversible derivatives of thyrotropin-releasing hormone (TRH) with increased lipophilicity and resistance to cleavage by the TRH-specific serum enzyme. *Pharm Res* 7: 885–892

39 Bundgaard H, Rasmussen, GJ (1991) Prodrugs of peptides. 9. Bioreversible N-alpha-hydroxyalkylation of the peptide bond to effect protection against carboxypeptidases or other proteolytic enzymes. *Pharm Res* 8: 313–322

40 Klixbüll U, Bundgaard H (1984) Prodrugs as drug delivery systems. 30. 4-Imidazolidinone as potential bioreversible derivatives for the alpha-aminoamide moiety in peptides. *Int J Pharm* 20: 273–284

41 Hughes JAH, Smith TV (1975) Identification of two related pentapeptides from the brain with potent opiate agonist activity. *Nature* 258: 577–579

42 Malik JB, Goldstein JM (1977) Analgesic activity of enkephalins following intracerebral administration in the rats. *Life Sci* 20: 827–832

43 Hambrook JM, Morgan BA, Rance MJ, Smith CF (1976) Mode of deactivation of the enkephalins by rat and human plasma and rat brain homogenate. *Nature* 262: 782–783

44 Tsuzuki N, Hama T, Hibi T, Konishi R, Futaki S, Kitagawa K (1991) Adamantane as a brain-directed drug carrier for poorly absorbed drug – antinociceptive effects of [D-Ala2]Leu-

enkephalin derivatives conjugated with the 1-adamantane moiety. *Biochem Pharmacol* 41: R5–R8

45   Kitagawa K, Mizobuchi N, Hama T, Hibi T, Konishi R, Futaki S (1997) Synthesis and antinociceptive activity of [D-Ala(2)]Leu-enkephalin derivatives conjugated with the adamantane moiety. *Chem Pharm Bull* 45: 1782–1787

46   Fredholt K, Adrian C, Just L, Larsen DH, Weng S, Moss B, Friis GJ (2000) Chemical and enzymatic stability as well as transport properties of a Leu-enkephalin analogue and ester prodrugs thereof. *J Control Rel* 63: 261–273

47   Prokai L (2002) Central nervous system activity of thyrotropin- releasing hormone and its analogues. In: EM Jucker (ed): *Progress in Drug Research*, Vol. 59. Birkhäuser, Basel, 133–169

48   Schally AV, Bowers CY, Redding TW, Barrett JF (1966) Isolation of thyrotropin releasing factor (TRF) from porcine hypothalamus. *Biochem Biophys Res Commun* 25: 165–169

49   Böler J, Enzmann F, Folkers K, Bowes CY, Schally, AV (1969) The identity of chemical and hormonal properties of the thyrotropin releasing hormone and pyroglutamyl-histidyl-proline amide. *Biochem Biophys Res Commun* 37: 705–710

50   Rondeel JMM, Klootwijk W, Linkels E, van de Greef WJ, Visser T (1994) Neural differentiation of the human neuroblastoma cell-line IMR32 induces production of a thyrotropin-releasing hormone-like peptide. *Brain Res* 665: 262–268

51   Ong SW, Liu HL, Pidgeon C. (1996) Immobilized-artificial-membrane chromatography: measurements of membrane partition coefficient and predicting drug membrane permeability. *J Chromatogr* A 728: 113–128

52   Prokai L, Zharikova AD, Janáky T, Li X, Braddy AC, Perjési P, Matveeva L, Powell DH, Prokai-Tatrai K (2001) Integration of mass spectrometry into early-phase discovery and development of central nervous system agents. *J Mass Spectrom* 36: 1211–1219

53   Patel ZD, McKinley BD, Davis TP, Porreca F, Yamamura HI, Hruby VJ (1997) peptide targeting and delivery across the blood-brin barrier utilizing synthetic triglyceride esters: design, synthesis and bioactivity. *Bioconj Chem* 8: 434–441

54   Rasmussen GJ, Bundgaard H (1991) Prodrugs of peptides. 15. 4-Imidazolidinone prodrug derivatives of enkephalins to prevent aminopeptidase-catalyzed metabolism in plasma and adsorptive mucosae. *Int J Pharm* 76: 113–122.

55   Lund L, Bak A, Friis GJ, Hovgaard L, Christrup LL (1998) The enzymatic degradation and transport of leucine-enkephalin and 4-imidazolidinone enkephalin prodrugs at the blood-brain barrier. *Int J Pharm* 172: 97–101.

56   Bak A, Fich M, Larsen BD, Frokjaer S, Friis GJ (1999) N-terminal 4-imidazolidinone prodrugs of Leu-enkephalin: synthesis, chemical and enzymatic stability studies. *Eur J Pharm Sci* 7: 317–323

57   Bundgaard H, Moss J (1989) Prodrugs of peptides. IV. Bioreversible derivatization of the pyroglutamyl group by N-acylation and N-aminomethylation to effect protection against pyroglutamyl aminopeptidase. *J Pharm Sci* 78: 122–126

58   Moss J, Bundgaard H (1992) Prodrug peptides. 19. Protection of the pyroglutamyl residue against pyroglutamyl aminopepitidase by N-acyloxymethylation and other means. *Acta Pharm Nord* 4: 301–308

59   Moss J, Buur A, Bundgaard H (1990) Prodrugs of peptides. 8. *In vitro* study of intestinal metabolism and penetration of thyrotropin-releasing hormone (TRH) and its prodrugs. *Int J Pharm* 66: 183–191

60   Moss J, Bundgaard H (1991) Prodrugs of peptides. 12. Bioreversible derivatization of thy-

rotropin-releasing-hormone (TRH) by N-phtalidylation of its imidazole moiety. *Int J Pharm* 74: 67–75

61   Muranishi S, Sakai A, Yamada K, Murakami M, Takada K, Kiso Y (1991) Lipophilic peptides: synthesis of lauroyl thyrotropin-releasing hormone, and its biological activity. *Pharm Res* 8: 649–652

62   Yamada K, Murakami M, Yamamoto A, Takada K, Muranishi S (1992) Improvement of intestinal absorption of thyrotropin-releasing hormone by chemical modification with lauric acid. *J Pharm Pharmacol* 44: 717–721

63   Greene DL, Hau VS, Abbruscato TJ, Bartosz H, Misicka A, Lipkowski AW, Hom S, Gillespie TJ, Hruby VJ (1996) Enkephalin analog prodrugs: Assessment of *in vitro* conversion, enzyme cleavage characterization and blood-brain barrier permeability. *J Exp Ther* 277: 1366– 1375

64   Lipkowski AW, Misicka A, Hosohata K, Davis P, Yamamura HI, Porreca F, Hruby VJ (2001) Biological properties of Phe(0)-opioid peptide analogues. *Life Sci* 68: 969–972

65   Bak A, Siahaan TJ, Gudmundsson OS, Gangwar S, Friis GJ, Borchardt RT (1999) Synthesis and evaluation of the physicochemical properties of esterase sensitive cyclic prodrugs of opioid peptides using an (acyloxy)alkoxy linker. *J Peptide Res* 53: 393–402

66   Wang B, Nimkar K, Wang W, Zhang H, Shan D, Gudmundsson O, Gangwar S, Siahaan T, Borchardt RT (1999) Synthesis, and evaluation of the physichochemical properties of esterase-sensitive cyclic prodrugs of opioid peptides using coumarinic acid, and phenylpropionic acid linkers. *J Peptide Res* 53: 370–382

67   Ouyang H, Borchardt RT, Siahaan TJ, Vander Velde D (2002) Synthesis and conformational analysis of a coumarinic acid-based cyclic prodrug of opioid peptide with modified sensitivity to esterase-catalyzed bioconversion. *J Peptide Res* 59: 183–195

68   Liao Y, Wang B (1999) Substituted coumarins as esterase-sensitive prodrug moieties with improves release rates. *Bioorg Med Chem Lett* 9: 1795–1800

69   Hui OY, Tang FX, Siahaan TJ, Borchardt RT (2002) A modified coumarinic acid-based cyclic prodrug of an opioid peptide: its enzymatic and chemical stability and cell permeation characteristics. *Pharm Res* 19: 794–801

70   Yang JZ, Chen W, Borchardt RT (2002) *In vitro* stability and *in vivo* pharmacokinetic studies of a model opioid peptide, H-Tyr-d-Ala-Gly-Phe-d-Leu-OH (DADLE), and its cyclic prodrugs. *J Pharm Exp Ther* 303: 840–848

71   Shan D, Nicolaou MG, Borchardt RT, Wang B (1997) Prodrug strategies based on intratmolecular cyclization reactions. *J Pharm Sci* 56: 765–767

72   Wang B, Gangwar S, Pauletti GM, Siahaan TJ, Borchardt RT (1997) Synthesis of a novel esterase-sensitive cyclic prodrug system for peptides that utilizes a "trimethyl"-lock-facilitated lactonization reaction. *J Org Chem* 62: 1363–1367

73   Gudmundsson OS, Jois SD, Vander Velde DG, Shiahaan TJ, Wang B, Borchardt RT (1999) The effect of conformation on the membrane permeation of coumarinic acid-and phenylpropionic acid –based cyclic prodrugs of opioids. *J Pept Res* 53: 383–392

74   Chen W, Yang JZ, Andersen R, Nielsen LH, Borchardt RT (2002) Evaluation of the permeation characteristics of a model opioid peptide, H-Tyr-D-Ala-Gly-Phe-D-Leu-OH (DADLE), and its cyclic prodrugs across the blood-brain barrier using an *in situ* perfused rat brain model. *J Pharmacol Exp Ther* 303: 849–857

75   Hoek J, Rydstrom J (1988) Physiological roles of nicotinamide nucleotide transhydrogenase. *J Biochem* 254: 1–10

76   Polgar L (2002) Structure-function of prolyl oligopeptidase and its role in neurological disorders. *Curr Med Chem CNS Agents* 2: 251–257

77   Bradbury A, Finnie M, Smith D (1982) Mechanism of C-terminal amide formation of pituitary enzyme. *Nature* 298: 686–688

78   Husain I, Tate SS (1983) Formation of the COOH-terminal amide group of thyrotropin-releasing-factor. *FEBS Lett* 152: 277–281

79   Prokai L, Prokai-Tatrai K, Bodor N (2000) Targeting drugs to the brain by chemical delivery systems. *Med Res Reviews* 20: 367–415

80   Walter R, Simmons WH, Yosimoto T (1980) Proline specific endo- and exopeptidase. *Mol Cell Biochem* 30: 111–1277

81   Prokai-Tatrai K, Prokai L, Bodor N (1996) Brain-targeted delivery of a Leu-enkephalin analogue by retrometabolic design. *J Med Chem* 39: 4775–4782

82   Szirtes TL, Kisfaludi L, Palosi E, Szporny L (1984) Synthesis of thyrotropin-releasing hormone analogues. 1. Complete dissociation of central nervous system effect from thyrotropin-releasing activity. *J Med Chem* 27: 741–745

83   Jackson I (1989) Controversies in TRH biosynthesis and strategies towards the identification of a TRH precursor. *Ann NY Acad Sci* 553: 71–75

84   Kahns AH, Bundgaard H (1991) Prodrugs of peptides. 13. Stabilization of peptide amides against α-chymotrypsin by the prodrug approach. *Pharm Res* 12: 1533–1538

85   Bundgaard H, Kahns AH (1991) Chemical stability and plasma-catalyzed dealkylation of peptidyl-α-hydroxyglycine derivatives-intermediates in peptide α-amidation. *Peptides* 12: 745–748

86   Fischer W, Spiess J (1987). Identification of a mammalian glutaminyl cyclase converting glutaminyl into pyroglutaminyl peptides. *Proc Natl Acad Sci* 84: 3628–3632

87   Busby WH Jr, Quackenbush GE, Humm J, Youngblood W, Kizer JS (1987) An enzyme(s) that converts glutaminyl-peptides into pyroglutamyl peptides. Presence in pituitary, brain, adrenal medulla and lymphocytes. *J Biol Chem* 262: 8532–8536

88   Prokai L, Ouyang X, Wu WM, Bodor N (1994) Chemical delivery system to transport a pyroglutamyl peptide amide to the central nervous system. *J Am Chem Soc* 116: 2643–2644

89   Prokai L, Prokai-Tatrai K, Ouyang X, Kim HS, Wu WM, Zharikova AD, Bodor N (1999) Metabolism-based brain-targeting system for a thyrotropin-releasing hormone analogue. *J Med Chem* 42: 4563–4571

90   Ishikura T, Senou T, Ishihara H, Kato T, Ito T (1995) Drug delivery to the brain. DOPA prodrugs based on a ring-closure reaction to quaternary thiazolium compounds. *Int J Pharm* 116: 51–63

91   Prokai L, Zharikova AD (2002) Neuropharmacodynamic evaluation of the centrally active thyrotropin-releasing hormone analogue [Leu^2]TRH and its chemical brain-targeting system. *Brain Res* 952: 268–274

92   Prokai-Tatrai K, Perjesi P, Zharikova AD, Li X, Prokai L (2002) Design, synthesis and biological evaluation of novel, centrally-acting thyrotropin-releasing hormone analogues. *Bioorg Med Chem Lett* 12: 2171–2174

93   Prokai L, Prokai-Tatrai K, Zharikova AD, Nguyen V, Stevens SM Jr (2003) Centrally-acting and metabolically stable thyrotropin-releasing hormone analogues by replacement of histidine with substituted pyridinium. *J Med Chem; in press*

Progress in Drug Research, Vol. 61
(L. Prokai and K. Prokai-Tatrai, Eds.)
©2003 Birkhäuser Verlag, Basel (Switzerland)

CNS-delivery via conjugation to biological carriers: physiological-based approaches

By Suresh P. Vyas

Drug Delivery Research Laboratory
Department of Pharmaceutical
Sciences
Dr. Harisingh Gour University
Sagar (M.P.) 470 003
India

Suresh P. Vyas

has nearly 22 years of research and teaching experience at both U.G. and P.G. levels. He has worked as Manager, R & D (New Drug Delivery) in Lupin Laboratories, Aurangabad for nearly one year. He is a leading scientist and well-known academician of international reputation. Prof. Vyas did his postdoctoral research at the School of Pharmacy, University of London (UK). He has supervised 18 Ph.D. and more than 60 M. Pharm. research projects. He has over 150 research publications to his credit published in international and national journals. He has been awarded Pt. Lajja Shankar Jha Award for excellence in research in Engineering and Technology by the Govt. of Madhya Pradesh. He is a recipient of Motan Devi Dandiya Oration Award, 1998 and 2001 and has been honoured with Best Pharmacy Teacher of the Year, 2001 Award by APTI. He has delivered invited lectures, keynote addresses and chaired many sessions in several National and International Conferences and Symposia in India and abroad. He has written many invited review articles in Advanced Drug Delivery Reviews, Critical Reviews™ in Therapeutic Drug Carrier Systems, Pharmaceutical Research *and* International Journal of Medicine *and written many chapters in national and international books. His research interest extends from novel drug delivery systems to process development, molecular basis of drug targeting, biopharmaceuticals, new drug delivery systems and molecular immunology. He has authored five books, entitled* Pharmaceutical Biotechnology, Advances in Liposomal Therapeutics, Controlled Drug Delivery: Concepts to Advances, Targeted and Controlled Drug Delivery: Novel Carrier Strategies *and* Methods in Biotechnology and Bioengineering.

Contents

Key words

Blood brain barrier, cationized antibodies, chimeric peptide, central nervous system, drug delivery, liposomes, nanoparticles, peptide nucleic acid.

Glossary of abbreviations

Ab, antibody; ADCC, antibody dependent cellular cytotoxicity; APP, amyloid precursor protein; AUC, area under curve; BBB, blood-brain barrier; BDNF, brain derived neurotrophic factor; CNS, central nervous system; DTT, dithiothreitol; HIV, human immunodeficiency virus; mAb, monoclonal antibody; NGF, nerve growth factor; NHS, N-hydroxy-succinimido; NMDA, N-methyl-D-aspartate; PNA, peptide nucleic acids; PEG, polyethylene glycol; P-gp, P-glycoprotein; RES, reticuloendothelial system; Tf, transferrin; TfR, anti-transferrin receptor; TNF, tumor necrosis factor; VIP, vasoactive intestinal peptide.

1   Introduction

The blood-brain barrier (BBB) is an insurmountable obstacle for a large number of bioactives including antiviral drugs, antineoplastic agents and central nervous system (CNS) active compounds like neuropeptides. Despite the formidable academic challenges of this problem, a great deal of research work has been conducted to explore possible role of molecular carrier complexes to improve transBBB transport of drugs. Recent advances in studies on BBB transport of xenobiotics vis-à-vis of nutrients and neuroactive agents have vividly transformed the classical concept of the BBB. This chapter critically covers physiologic based strategies, which employ pseudonutrients, cationic antibodies, chimeric peptides, recombinant protein(s) and peptides transport system for vectoring of impervious biomolecules across the BBB.

2   Nutrient transport across the BBB

Several transport systems for nutrients and endogenous biologicals exist in the brain capillary endothelial cells, which constitute the BBB [1, 2]. These transport systems include (1) the hexose transport system for glucose and mannose, (2) the neutral amino acid transport system for phenylalanine, leucine and other neutral amino acids, (3) the acidic amino acid transport system for glutamic acid and aspartic acid, (4) the basic amino acid transport system for lysine and arginine, (5) the β-amino acid transport system for β-alanine and taurine, (6) the monocarboxylic acid transport system for lactic acid and short-chain fatty acids, such as acetate and propionate, (7) the choline transport system for choline and thiamine, (8) the amino transport system for mepyramine, (9) the nucleoside transport system for purine bases

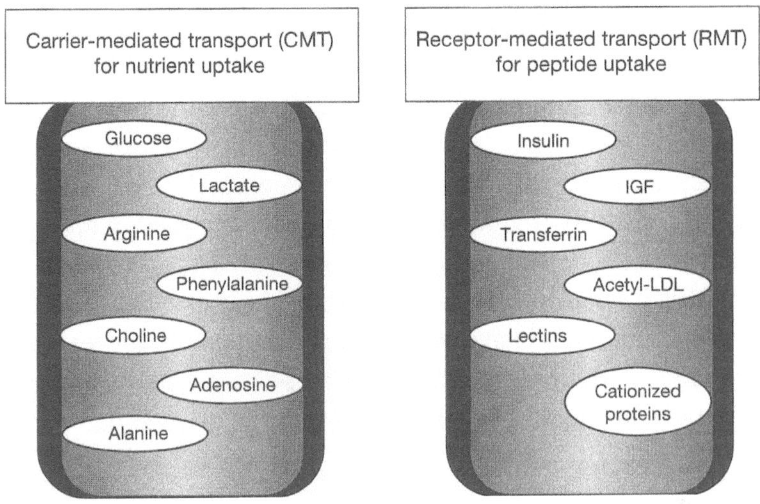

Figure 1.
Carrier-mediated and receptor-mediated uptake at the blood brain barrier.

such as adenine and guanine excluding pyrimidine bases, and (10) the peptide transport system for small peptides such as vasopressin.

Differences in the affinity and the maximal transport activity among these transport systems can be used selectively for the design of strategies to control or retard the delivery of drugs into the brain. Different transport systems identified associated with the BBB are schematically shown in Figure 1.

3 Antibodies for brain delivery

3.1 Cationized antibodies

Physiologically based strategies utilize intrinsic transport mechanisms that operate for transport of macromolecules across BBB. These transportation mechanisms include receptor-mediated transcytosis and absorptive transcytosis. The absorptive transcytosis operates particularly in the case of cationic proteins. Cationic antibodies have exhibited higher affinity and receptor binding. Thus, cationization of antibodies could technically be a more realistic and a more practically relevant approach for targeted drug delivery to the brain. Additionally, chemical modifications such as radiolabelling can influence the pharmacokinetics of cationized antibodies. The plasma membrane of brain endothelial cells, like other cell membranes, is negatively charged due to the anionic microdomains [3, 4]. This overall negatively charged surface on the luminal side of endothelium cells contributes to the barrier function. Ionic interactions between positively charged (cationic) proteins and negatively charged domains on cell membranes initiate absorptive mediated endocytosis. Polycationic proteins are known to penetrate the BBB using this mechanism [5]. Other proteins, i.e. horse radish peroxidase [6] and native albumin [7] have been conjugated to protamine for successful delivery to brain, where conjugated proteins permeate across BBB through absorptive mediated uptake by endothelial cells.

3.2   Genetically engineered antibodies

Antibodies exhibit multiple functions that make them key components of the mammalian humoral immune response and versatile therapeutic agents *per se* for a variety of applications. Advances in genetic engineering and customized protein expression systems have led to enormous progress in the development of immunoglobulins with defined novel functional properties. Recombinant technology can also be utilized for immunogenicity masking or alleviation. Recently, antibodies have been generated with novel properties such as dual binding specificities (bifunctional antibodies), i.e. combining antigen recognition site (domain) with molecules such as toxins or growth factors (antibody fusion proteins). However, all the recombinant antibodies must be evaluated for their specificity and affinity. Taken together, the accessibility to target antigen, effector functions and clearance determine the ultimate usefulness of recombinant antibodies. Moreover, safety, which depends in part on the host reaction to the foreign protein, its pharmaceutical composition, cross-reactivity with non-target antigens as well as uptake by inappropriate tissues may become the ultimate decisive consideration in the development of recombinant protein(s).

3.3   Chimeric and humanized antibodies

With the advent of hybridoma technology it is now possible to produce antibodies with defined specificity. However, monoclonal antibodies (mAbs) produced in rodents have shown immunogenicity and also they are rapidly cleared from blood circulation. To circumvent this limitation protein engineering has been judiciously employed to clone up murine mAbs as mouse/human chimeric antibodies either by grafting the antigen binding sites of rodent antibodies onto the variable framework regions of human antibodies [8, 9] or by joining human constant region domains to murine variable region domains [10, 11].

Chimeric antibodies are produced with predetermined combinations of light and heavy chain variable and constant region genes. These antibodies are amenable to isotype switching, domain substitution and mutagenesis. Chimeric antibodies are invaluable tools for the study of key Ab effector functions such as Fc receptor binding, Ab dependent cellular cytotoxicity (ADCC) and complement activation, which vary considerably among IgM, IgA and the four IgG isotypes [12, 13]. ADCC is an important *in vivo* operative mechanism through which a biosystem eliminates cellular targets such as tumor cells.

3.4 Antibodies with novel effector functions

The specificity of an Ab to an antigen endows them with a multitude of clinical application potentials. Modern techniques are available for producing functional Ab fragments. It is also possible to join Ab-binding specificities to a non-Ab effector sequence(s).

mAbs with catalytic properties that resemble enzymes were first produced in 1986 [14, 15]. Generation of antibodies capable of amide bond hydrolysis [16] have open up a new vista of abzymes and also generated an optimism *per se* that antibodies, which catalyzed therapeutically important reactions such as peptide hydrolysis could be successfully produced.

The combination of functions are well examplified by catalytic antibodies, i.e. abzyme, bifunctional with site specificity and effector functionality, adhesion immunoglobulins and non-immunoglobulin sequences substituted antibodies [16, 17]. The Ab combining specificity can be used to provide specific delivery of an associated biological activity. Joining tumor necrosis factor (TNF) to an anti-transferrin receptor (TfR) Ab [18] resulted in a fusion protein with TNF cytotoxic activity towards cell lines with the TfR. Such systems could be implicated as cytotoxic biomolecular complexes for their delivery to brain tumor.

3.5 Antibodies and fusion proteins for brain targeting

Brain delivery has always been a challenging task due to poor transport of water soluble drugs through the brain capillary endothelial wall, which constitutes the BBB. Brain capillary endothelial tight junctions impose a stern barrier against uptake of circulating solutes [19]. As previously described, in order to obtain required nutrients and factors from the blood, the BBB possesses some specific receptors, which transport compounds from circulating blood pool to the brain. Studies indicated that some compounds like insulin [20], transferrin [21] and insulin like growth factors 1 and 2 [22, 23] traverse the BBB *via* receptor-mediated transcytosis. Receptors for these molecules thus provide potential means to an access to the brain (Fig. 2).

Receptor specific antibodies offer an alternative stratagem for brain targeting. Immunoglobulins, as such, are excluded from the brain [24], however an Ab to the transferrin receptor (OX-26) has been shown to be dynamically transported into the brain parenchyma and therapeutically effective as a drug delivery vehicle [25]. OX-26 Ab has demonstrated an effective vehicle for the delivery of nerve growth factor (NGF) that otherwise fact to permeate BBB in therapeutically effective quantity [26].

4   Vector-mediated drug delivery

Vector-mediated drug delivery to the brain is essentially based on chimeric peptide technology, wherein transportable vectors such as cationized albumin or a receptor-specific mAb is conjugated to a therapeutic compound that under normal state of bioconditions is not transported through the BBB. The conjugation of drugs to transport vectors is facilitated by the use of avidin-biotin technology. Multiple classes of therapeutics have been delivered to the brain using chimeric peptide carriers, including peptide-based pharmaceuticals such as a vasoactive intestinal peptide or neurotrophins such as brain-derived neurotrophic factor, antisense therapeutics including peptide nucleic acid and small molecules incorporated within liposomes (Fig. 3).

4.1 Chimeric peptide model

Brain capillary endothelium possesses peptide receptors that physiologically mediate peptide transport through the BBB. This constitutes the platform for

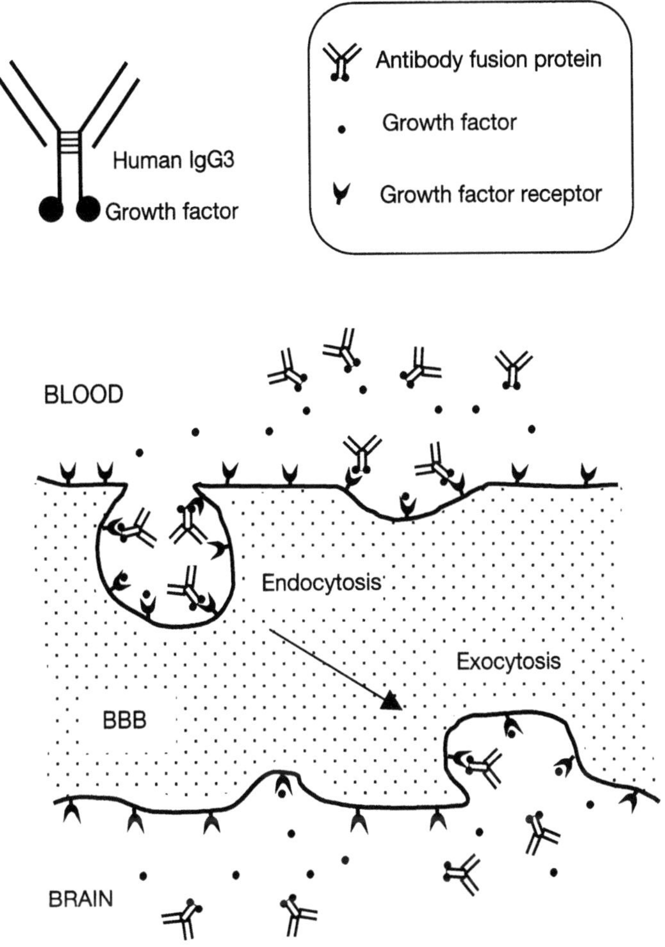

Figure 2.
Transport of molecules into the brain using the growth factor fusion proteins with antibody.

the chimeric peptide hypothesis, where drug delivery to the brain may be attained by the linking of drugs to a peptide or protein vector, which could traverse across the BBB into brain from blood by absorptive or receptor mediated transcytosis. The chimeric peptide system traverses the BBB involving four distinctive steps (Fig. 4).

The first step is receptor mediated endocytosis of the chimeric peptide at the luminal surface of BBB followed by endosomal movement across the 300 nm of endothelial space. The second step is exocytosis into brain interstitial

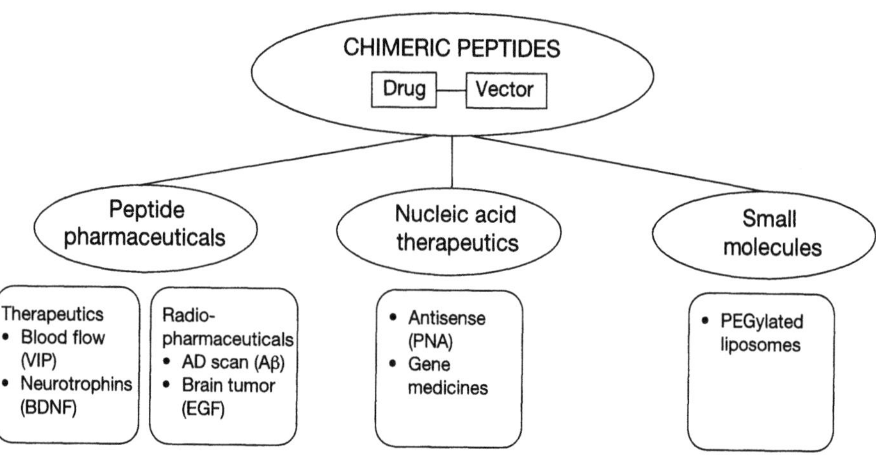

Figure 3

Versatility of chimeric peptide technology, based carriers for delivery of therapeutics, nucleic acids and small molecules.

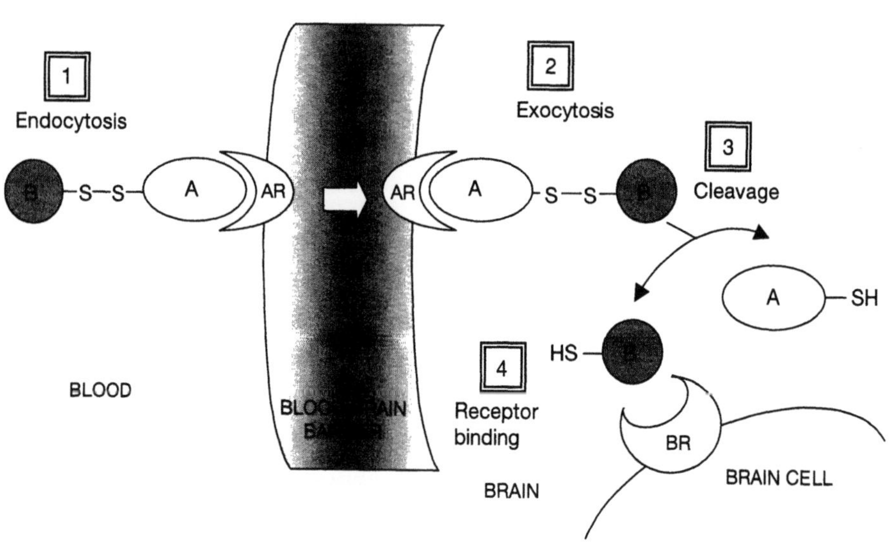

Figure 4.

Sequential steps for delivery of chimeric peptides through the blood-brain barrier.

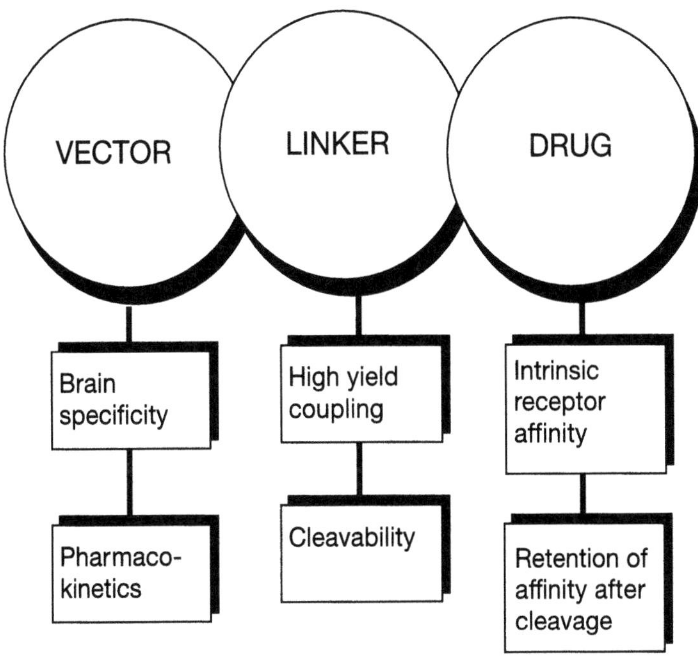

Figure 5.
Brain drug delivery using chimeric peptide technology involves three interdependent spheres: the vector sphere, the linker sphere and the drug sphere.

fluid. This is followed by cleavage of the disulfide bond liberating the therapeutic compound from the vector. The fourth and last step is binding of the therapeutic peptide to its receptors. The chimeric peptide technology is comprised of three inter-dependent spheres: the vector sphere, the linker sphere and the drug sphere (Fig. 5). Vector specificity for brain, vector pharmacokinetics, high yield coupling between vector and drug, cleavability of linker and intrinsic receptor affinity for drug following release from the transport vector are some important considerations, for efficient chimeric peptide design.

Efforts must be made to explore or design and develop some more efficient BBB transport vectors. There is also a need to develop more efficient linker strategies that could result in to high yield coupling of drug to therapeutic vector. Furthermore, additional attention must be concentrated on drug development sphere, wherein bioactivities could be monobiotinylated at the sites that are not critical to receptor binding.

4.2 Vector discovery

4.2.1 Cationized albumin

Cationized BSA cannot be used in humans since the cationization of heterologous proteins results in a marked increase in immunogenicity [27]. Enhanced uptake of cationized protein by antigen presenting cells accounts for increased immunogenicity of a cationized heterologous protein. However, studies suggested that cationized homologous proteins have minimal intrinsic immunogenicity and exhibit no toxic effects on continuous administration even up to eight weeks [28].

4.2.2 Protamine

Protamine is a cationic protein rich in arginine (50%) and is produced in high concentrations in spermatozoa [29]. High concentration, of arterial protamine causes biochemical blood-brain barrier disruption [30, 31]. Protamine has a dual effect on vascular permeability (Fig. 6). First, it causes a generalized increase in vascular permeability, particularly following carotid artery infusion in the absence of plasma proteins. This generalized increase in BBB permeability may enhance the transport of low molecular weight solutes like sucrose. Second, following intravenous administration, when it circulates in the presence of plasma proteins, a vectorial transfer of protamine-protein complexes takes place across the microvascular beds of brain.

The vectorial transfer of protamine-protein complexes is a result of the amphiphilic properties of protamine, i.e. the ability to bind both proteins like albumin or globulin on one end and negatively charged capillary endothelium surface on the other. This dual binding may subsequently trigger the absorptive transcytosis of the protamine protein complex.

4.2.3 Recombinant CD4

CD4 receptors are expressed on T4 lymphocytes, which have strong affinity for gp120 surface protein of the HIV [32]. Recombinant CD4, however lacks the transmembrane region hence it was initially proposed as circulatory receptor for sequestration of gp120 [33]. Since HIV may also harbor in the brain, localized delivery of antiviral therapeutics is required for effective

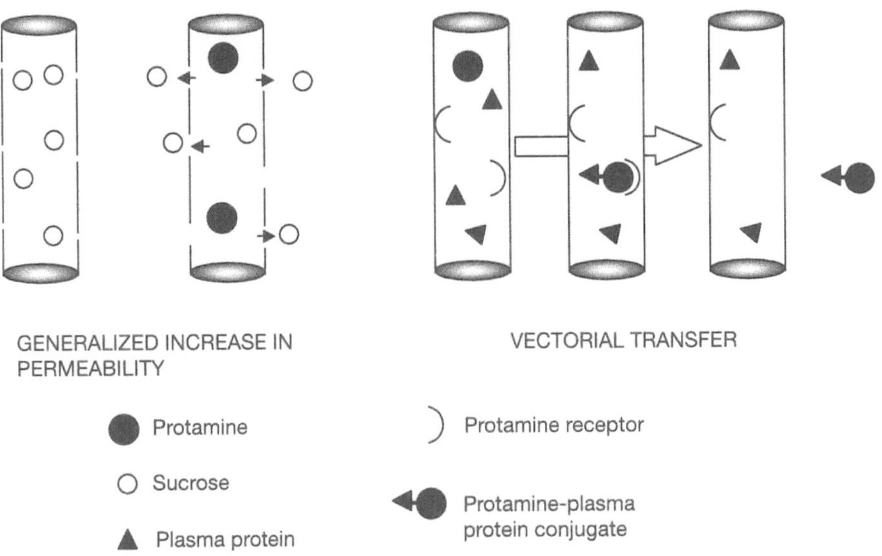

Figure 6.
Effect of protamine on vascular permeability.

treatment. Recombinant CD4 protein has been investigated for brain targeting of anti-AIDS drugs. Recombinant CD4 is essentially a cationic protein. It was therefore, anticipated that CD4 protein uptake by brain capillaries may be competitively inhibited by protamine. Other cationic proteins like cationized BSA or histone exhibit saturable binding to brain capillaries endothelium, however it is inhibited by protamine [34]. In contrast the brain capillary uptake of recombinant CD4 is not inhibited by protamine even at its blood level of 25 µg/ml [35]. This unexpected effect may be ascribed to the dual actions of protamine, whereby it inhibits CD4 binding to brain capillaries and simultaneously mediates the uptake of recombinant CD4 through the protamine/CD4 complexes (Fig. 7).

4.2.4 Anti-transferrin receptor antibody

It is well documented that the brain microvasculature possesses numerous transferrin receptors and these receptors functionally mediate the transcytosis of transferrin across the BBB [36]. Studies revealed that receptor mediated transcytosis of transferrin through the BBB is a saturable phe-

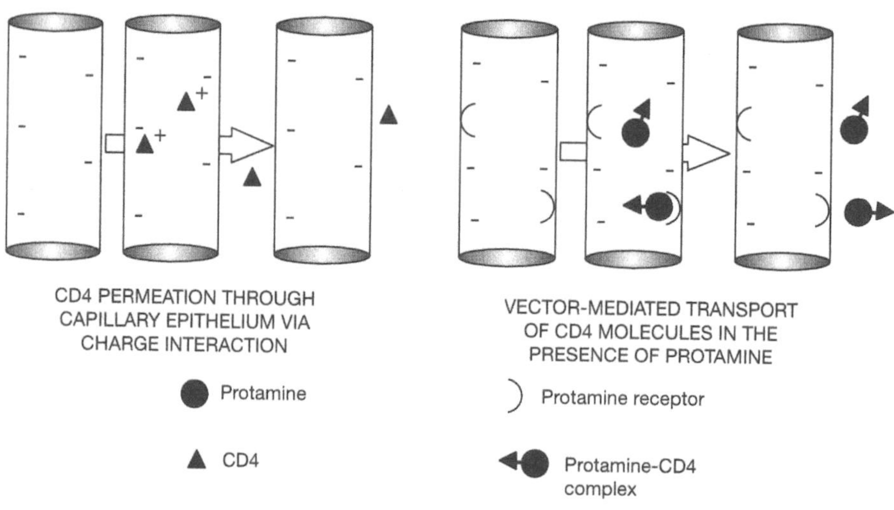

Figure 7.
CD4 protein transport across capillary epithelium and possible role of protamine.

nomenon that proceeds at rates virtually identical to the transcytosis of the OX26 Ab [37]. The OX26 Ab is a murine mAb produced against rat transferrin receptor. It binds to an extracellular projecting epitope on the receptor located distantly away from the binding site of transferrin. The OX26 Ab selectively binds to microvascular endothelium in brain and undergoes receptor-mediated transcytosis through the BBB [38, 39]. The OX26-mediated delivery to brain of biotinylated drugs is possible by preadministration of conjugates of OX26 and neutral avidin [40]. It has been observed that OX26/NLA (Ab:neutral avidin) conjugate crosses the BBB more efficiently compared to cHSA/NLA (cationic human serum albumin:neutral avidin) conjugate [41].

Studies have demonstrated that transferrin (Tf)-Ab fusion proteins possess TfR binding ability and can specifically be targeted to the brain through the TfR expressed in abundance on the BBB [42]. An important characteristic of the Tf Ab fusion proteins is that they retain their ability to bind antigen and can potentially be utilized either for secondary targeting within the brain or for non-covalent attachment and subsequent delivery of a therapeutic molecule, through their combining site. Depending on the Ab constructs, the Tf Ab fusion proteins can be targeted with or without associated effector functions. These proteins offer a potentially useful family of diagnostics and therapeutic reagents for sites rich in TfRs such as the brain.

4.2.5 Anti-insulin receptor antibody

Insulin receptors are present on brain capillary endothelium that mediate the transcytosis of circulating insulin and anti-insulin receptor mAbs across the BBB. Two different murine mAbs to the human insulin receptor have been produced and designated as mAb83-7 and mAb83-14 [43, 44]. It has been observed that mAb83-7 binds through an epitope within the amino acid region between 191-297 of the α-subunit and mAb83-14 binds an epitope within the amino acid region between 469-592 of the α-subunit of the human insulin receptor [44]. mAb83-14 selectively binds to the human brain microvessel to a much higher extent compared to mAb83-7 and approximately 70% of the former gets endocytosed. The avid binding of mAb83-14 to human brain capillaries is due to very high affinity of this binding interaction [45]. In order to establish transcytosis of mAb83-14 Ab across the BBB *in vivo* experimentation is mandatory. Studies have shown that the brain delivery of mAb83-14 in the Rhesus monkey is approximately eight fold higher than the brain delivery of OX26 antibodies in rat.

4.3 Linker development

4.3.1 Chemical cross-linking

Chemical cross-linking based methods could be used for conjugation of therapeutic compounds such as β-endorphin analogues to model vectors like cationized BSA [46]. There are numerous cross-linking reagents available specific to surface associated amino, carboxyl or thiol groups. However, conventional chemical cross-linking methods suffer from inherent limitation of low yield of the cross-linking reaction. The yield obtained was only 10-15%, where a β-endorphin analogue was chemically conjugated to cationized BSA [46]. However, to further increase the yield and binding efficiency avidin-biotin technology was used [47]. Chemical cross-linking is generally employed to synthesize a vector/avidin conjugate followed by its coupling to the monobiotinylated ligands. Surface carboxyl or amino groups of ligands

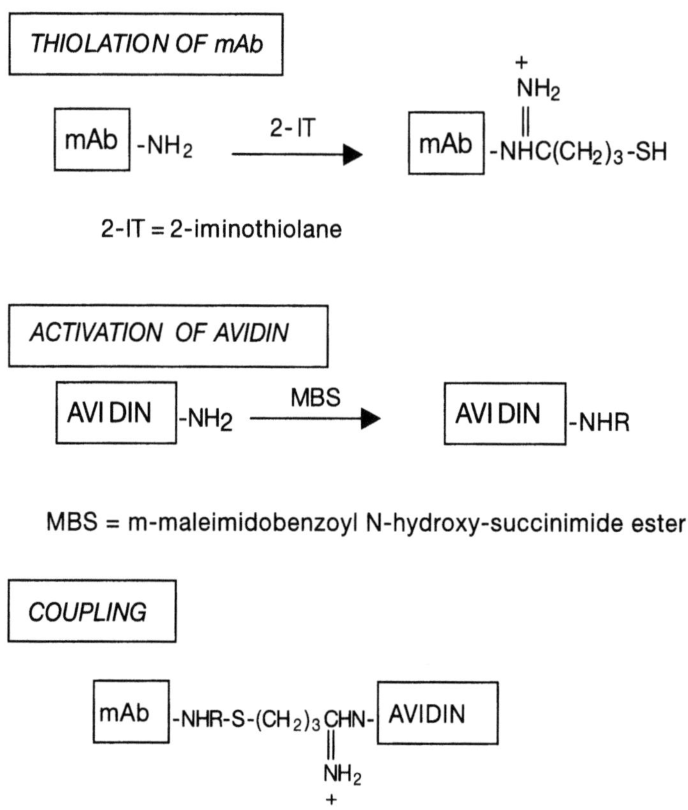

Figure 8.
Conjugation of monoclonal antibody and avidin *via* a thiol-ether linkage.

can be monobiotinylated using biotin analogues with a cleavable (e.g. disulfide) or non-cleavable (e.g. amide) linkage.

4.3.2 Avidin-biotin technology

Avidin is a homotetramer with monomeric units each having 120 amino acids. The entire tetramer is approximately 64 kDa [48]. Avidin is an avian cationic protein with an isoelectric point (pI) of 10. It is an extremely stable protein. It is not denatured by extreme pH or high concentrations of detergent. Avidin binds biotin with extremely high affinity with a k_D of approximately 10^{-15} M with a dissociation half-time of approximately 89 days [49].

Streptavidin is a bacterial homolog of avidin, which is non-glycosylated and is slightly acidic with a pI of 5–6 and possesses an amino acid homology with avidin to an extent of 38% [50]. Conjugation of avidin with mAb is schematically shown in Figure 8. The use of avidin-biotin interaction in drug delivery involves the single administration of a complex of vector/avidin fusion protein conjugated to a monobiotinylated therapeutic agent. The therapeutic agent must be monobiotinylated since high degree of biotinylation renders the high molecular weight aggregation susceptible to reticuloendothelial system sequestration and subsequent clearance.

Besides, avidin or neutral avidin Ab/avidin fusion proteins can be produced using genetic engineering, wherein the Fc portion of the IgG heavy chain is replaced with an avidin dimer. The use of Ab/avidin fusion proteins in humans must consider the potential antigenicity of the conjugate. However, the antigenicity of the mAb may be tailored through humanization of the murine Ab or with the use of fully human Ab [51].

5   Bioactive peptides

5.1   Vasoactive intestinal peptide (VIP)

Vasoactive intestinal peptide (VIP) is known to be a potent cerebrovasodilator after topical application to intracranial blood vessels [52], however after administration of VIP *via* the carotid artery, very low cerebrovasodilatation was produced [53]. This may be due to lack of BBB permeability of VIP after systemic administration. VIP transport across the BBB can be increased by using chimeric peptide technology. For BBB permeation enhancement a VIP analogue (VIPa) was designed for monobiotinylation which retained intrinsic biologic activity following its conjugation to the OX26/AV vector [54]. The OX26 transferrin receptor mAb (TfRMAb) is conjugated to avidin *via* a stable thiol-ether linkage. The conjugate is further conjugated to a biotin moiety attached to the VIP analogue *via* a disulfide bond (Fig. 9). It has been observed that an epitope on the transferrin receptor (TfR) binds the entire complex, which subsequently proceeds through BBB transcytosis of the biotinylated VIPa. Two criteria must be fulfilled during biotinylation of the VIP, monobiotinylation and retention of biologic activity after cleavage of the disulfide linkage. Studies revealed that when bioVIPa-avidin/OX26 was infused at a dose of 12 µg/kg peptide a 65% increase in

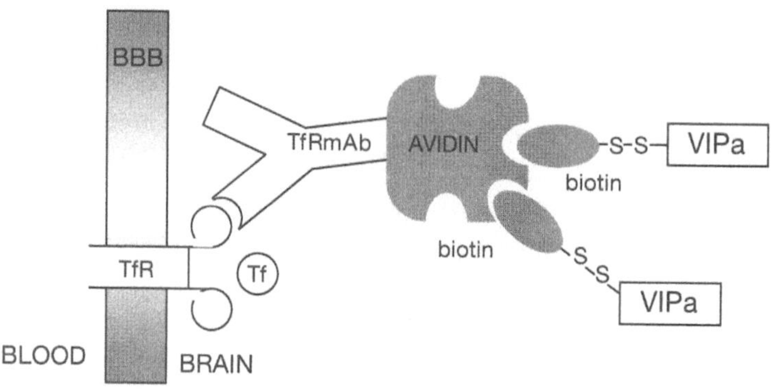

Figure 9.
Delivery of biotinylated-SS-vasoactive intestinal peptide analogue (VIPa) across the BBB through avidin transferrin receptor monoclonal antibody conjugate (mAb).

hemispheric brain blood flow was measured [54]. This increase in cerebral blood flow caused by the VIP analogue coupled to BBB delivery vector is attributed to release of the VIP analogue to vascular smooth muscle cells located in brain. It is well established that these smooth muscle cells are richly innervated with VIP nerve endings [55], as VIP is a major cortical vasodilator [56]. VIP alone could not increase cerebral blood flow, which is in accordance with other studies showing that in the absence of a specific brain drug delivery system VIP is not significantly transported across the BBB.

5.2   [D-Arg^2, Lys^4] dermorphin analogue (DALDA)

DALDA is a μ-opioid peptide receptor agonist. DALDA was biotinylated at the Lys^4 position with N-hydroxy-succinimide (NHS)-SS-biotin and the so formed bioDALDA was treated with dithiothreitol (DTT) to produce desbio-DALDA, which contains a mercaptopropionate group at Lys^4 position [57]. Studies conducted to establish the affinity of molecules to receptors showed that the affinity of the desbioDALDA to the μ-receptor was comparable to that of DALDA. A significant loss in affinity for the μ-receptor was recorded, when bioDALDA was conjugated to avidin without cleavage of the disulfide bond [57].

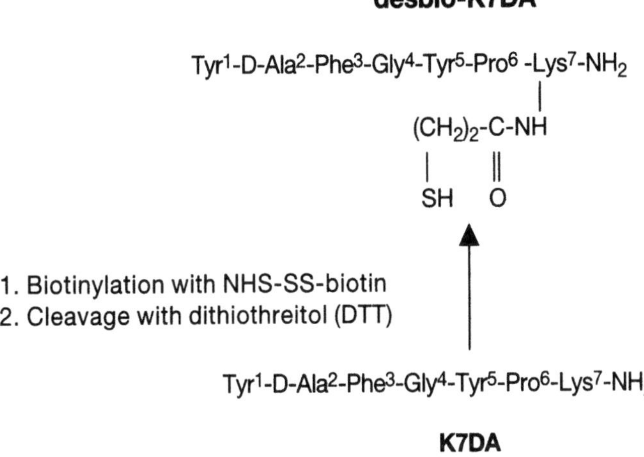

Figure 10.
Reaction involved in biotinylation and cleavage of K7DA.

5.3 [Lys^7] dermorphin analogue (K7DA)

K7DA is a μ-opioid peptide receptor specific analogue with structural features similar to DALDA, however it has an approximately 4-fold higher affinity for the μ-opioid peptide receptor as compared to DALDA [58]. K7DA was biotinylated with both NHS-SS-biotin to form bio-SS-K7DA and was biotinylated with biotin-XX-NHS to form bio-XX-K7DA, where XX represents bis-amino hexanoyl spacer arm and NHS, N-hydroxysuccinimide. The bio-SS-K7DA was cleaved using DTT to produce desbioK7DA (Fig. 10). Studies suggested that K7DA, like DALDA could be biotinylated and cleaved with DTT whilst it still retains μ-opioid receptor affinity. Studies also indicated that although bio-XX-K7DA retains high affinity for the μ-opioid receptor but this affinity is abolished once bio-XX-K7DA is linked to NLA-OX26 vector [58] suggesting that in case of K7DA a cleavable or disulfide linker is essential rather than a non-cleavable or amide linker.

5.4 Brain derived neurotrophic factor (BDNF)

BDNF is a neurotrophin, and is a member of nerve growth factor (NGF) family. Other members of this family include NGF, neurotrophin-3 and neu-

Table 1.
Pharmacokinetic problems that limit brain drug delivery

Pharmacokinetic problem	Example	Solution of problem
Rapid plasma clearance of vector	Histone	Use different vector
Rapid plasma clearance of avidin/vector conjugate	OX26/AV	Use neutral avidin
Rapid plasma clearance of neurotrophin	BDNF	PEGylation
Rapid peptidase inactivation of peptide	Endorphin/amino-peptidase M	[D-Ala^2] substitution

rotrophin 4/5. *In vivo* studies confirm the avid binding of BDNF to the brain capillaries inner wall, but indicate that there is no measurable transport of the neurotrophin the brain [59, 60]. It has been demonstrated that simply conjugating drugs to BBB specific vectors, without optimizing plasma pharmacokinetics, may reduce the ultimate delivery of the neurotrophin to brain. Neurotrophin BBB permeability surface area product increases by its conjugation to a transport vector, whereas plasma area under curve (AUC) can be enhanced particularly using polyethylene glycol (PEG) conjugation (Tab. 1).

6   Antisense drug delivery

Antisense drugs are potential therapeutics used in the treatment of cancer, viral infection and other disorders of brain and peripheral tissues. However, the *in vivo* efficacy of these oligomers is limited due to serum and cellular nuclease-mediated degradation and minimal transcapillary transport or cellular uptake. Antisense oligodeoxynucleotides (ODN) are molecules with high degrees of specificity, since these molecules react in a sequence specific mechanism with target messenger RNA (mRNA) molecules in the cell cytosol. Phosphodiester antisense oligodeoxynucleotides (PO-ODNs) were the first generation antisense ODNs with dual sites of action; RNA degradation *via* activation of RNase and arrest of RNA translation by formation of DNA-RNA duplexes. PO-ODNs, however, are non-effective pharmaceuticals *in vivo* because these molecules are rapidly degraded by nucleases. The second generation ODNs constitute phosphothioate PS-ODNs with an advantage of nuclease resistance. However, PS-ODNs turned out to be inhibitors of RNase H [61] and are bound to multiple cellular proteins. These antisense molecules produce some serious side effects, even leading to death [62].

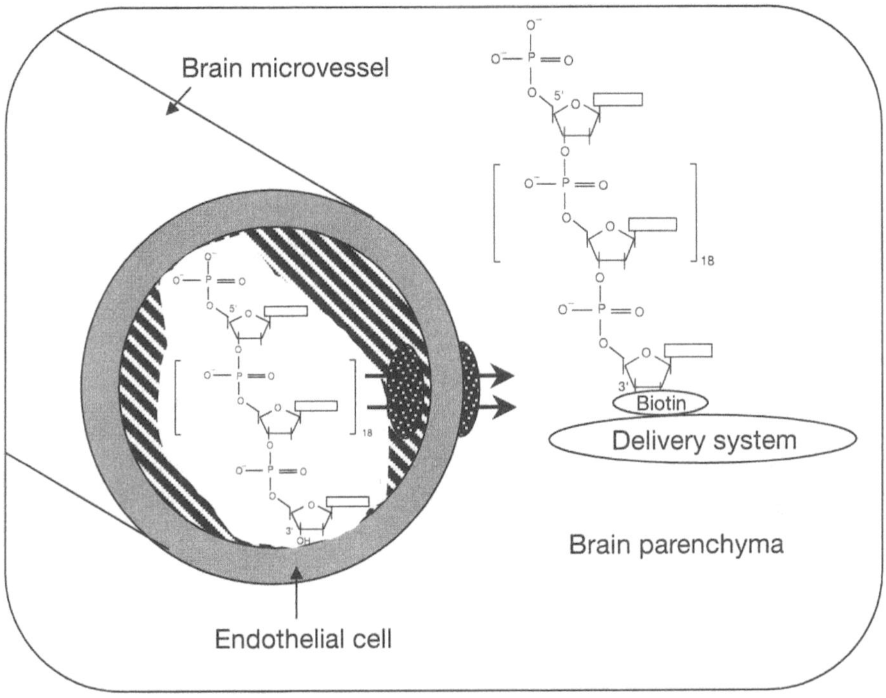

Figure 11.
Oligodeoxynucleotide transport through the BBB.

A series of studies has examined the extent to which antisense ODNs may be transported through the BBB through vector mediated drug delivery systems (Fig. 11). Colloidal delivery systems such as liposomes and nanoparticles are not effective as these are mainly taken up by cells lining the reticulo-endothelial system. Surface modifications of colloidal carriers, such as anchoring of PEG and Ab attachment, enable them to transport antisense oligodeoxynucleotide across the BBB. Biotinylated antisense ODN can be efficiently delivered to the liver and kidney by simply conjugating to avidin. Although biotinylation of the 3'-terminus of phosphodiester ODN has been shown to protect the oligomer completely against serum and cellular nucleases. Similarly, conjugation to the delivery vector cationic human serum albumin-avidin, markedly increases the uptake (84-fold) in peripheral blood lymphocytes.

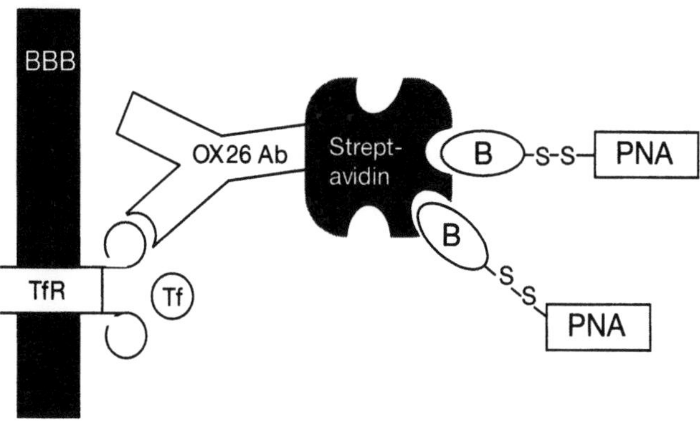

Figure 12
Vector-mediated delivery of a peptide nucleic acid through the BBB by biotinylation of the PNA *via* a disulfide bridge and conjugation to a complex of the OX26 monoclonal antibody and streptavidin.

7 Peptide nucleic acids (PNAs)

PNAs are molecules with antisense and antigene properties with a polypeptide backbone [63]. Studies revealed that both phosphodiester (PO) oligodeoxy-nucleotides (ODN) and phosphorothioate (PS)-ODN have undesirable pharmacokinetics [64]. However, it has been observed that biotinylated PNAs may be effectively conjugated to the OX26/SA vector and this allows for significant brain delivery of PNS [65].

Studies demonstrated that in the absence of a transcellular delivery system, the transport of PNA through the BBB is almost negligible. This finding substantiates the results of other experiments showing that PNAs do not possess biologic effects on cells unless they are actually directly injected into the cell cytoplasm [66]. However, it has been established that biotinylated PNAs may be delivered to brain by coupling these antisense or antigen compounds to vector based delivery systems that are transported through the BBB [65] (Fig. 12).

7.1 Diagnosis of Alzheimer's disease

Diagnosis of Alzheimer's disease has always been difficult due to the lack of a means to visualize and target β-amyloid plaques in the brains of affected patients. No method is available for detecting amyloid plaques in living subjects. However, many methods are available for detection by staining post-mortem brain tissue. Anti-β-amyloid antibodies have been proposed as a highly specific probe to monitor amyloid plaque formation in living patients [67]. Studies showed that intranasal administration of a filamentous phage as a delivery vector of anti-β-amyloid Ab fragment into Alzheimer's amyloid precursor protein (APP) transgenic mice enables *in vivo* targeting of β-amyloid plaques. Thioflavin-S and fluorescent-labeled antiphage antibodies were used for visualization of plaques in the olfactory bulb and the hippocampus region. The genetically engineered filamentous bacteriophages offer advantage, over other mammalian vectors by being efficient and non-toxic delivery vectors to the brain. This diagnostic procedure for detection of Alzheimer's disease provides a useful diagnostic tool for imaging Aβ deposits in living patients.

Another strategy for amyloid imaging utilized monobiotinylated ^{125}I-Ab1-40 conjugated to a BBB drug delivery system comprised of 83-14 mAb to the human insulin receptor, which is tagged with streptavidin [68]. This study described a methodology for BBB drug delivery and brain targeting of peptide radiopharmaceuticals for imaging amyloid or other disorders. Different physiological approaches for CNS delivery of drugs and bioactives are enlisted in Table 2.

8   Colloidal carrier systems for brain drug delivery

8.1   Liposomes

Several studies confirm that liposomes, even small unilamellar vesicles do not undergo transport through the BBB [72, 73]. Liposomal delivery across the BBB is only possible after suitable makeup of the liposomal surface so that the liposomes can bypass the reticuloendothelial system (RES) and remain intact. As shown in Table 3 various strategies have been devised for brain delivery, however almost every strategy involves an approximate 1:1 stoichiometry of vector to drug. The carrying capacity of vector could be greatly improved if the drug is entrapped in liposomes that are further linked to the BBB transport vector. Liposomes are rapidly removed after intravenous injection by the RES cells. Both RES bypass and BBB uptake can be addressed by PEGylation and utilizing chimeric peptide technology [74]. In this strategy, a novel bifunctional PEG2000 derivative that contains a maleimide at one

Table 2.
Summary of various physiological approaches for brain targeting.

S. No.	Bioactive	Remarks	Refs.
Vector mediated delivery			
1	VIP	Significant increase in transport across the BBB	[53, 54]
2	DALDA	Retains biological activity after disulfide bond cleavage, used for designing monobiotinylation strategies	[57]
3	K7DA	For biological activity of antibody conjugate cleavable linker is required	[58]
4	BDNF	Increase in BBB transport	[60]
5	PNA	Biotinylated PNA may be transported across the BBB	[65]
6	Anti-β amyloid antibodies	Diagnosis of Alzheimer's disease	[67]
7	Peptidal radio-pharmaceuticals	Imaging of amyloid for Alzheimer's disease detection	[68]
8	Radiolabeled EGF, PNA	Imaging of brain tumors	[69, 70]
9	Genes encoding luci-ferase or β-galactosidase	Widespread gene expression in the blood achieved	[71]
Colloidal carrier systems			
Liposomes			
10	Daunomycin	PEGylated immunoliposomes are capable of receptor mediated transport through the BBB	[74]
11	β-galactoside	Successful internalization through the BBB	[75]
Nanoparticles			
12	Dalargin, kytorphin, loperamide, tubocurarine, MRZ 2/576, doxorubicin	Polysorbate coated nanoparticles mimic LDL particles, thereby interact with LDL receptors leading to uptake by endothelial cells and release of drug into the brain interior	[77–81]

end, for attachment to a thiolated mAb and a distearoyl phosphatidyl ethanolamine moiety at the other end for incorporation into the liposome surface was used to construct the PEGylated immunoliposomes for the delivery of daunomycin (Fig. 13). The combined use of PEGylation technology, liposome technology and chimeric peptide technology results in the development of PEGylated immunoliposomes that are capable of receptor mediated transport through the BBB. In fact a single 100 nm liposome can entrap more than 10,000 small molecules of drug, the use of liposome technology could in combination with chimeric peptide technology exponentially increase the carrying capacity of the BBB transport vector. This made it possible to administer relatively large systemic doses of drug without significantly increasing the mAb transport vector.

Table 3.
Various strategies used for linking drugs to transport vectors.

Class	Target AA	Agent	Linkage	Cleavability
Chemical	Lys	MBS	Thio-ether (-S-)	No
Lys	Traut's			
Lys	SPDP	Disulfide (-SS-)	Yes	
Lys	Traut's			
Avidin-biotin	Lys	NHS-SS-biotin	Disulfide	Yes
Lys	NHS-XX-biotin	Amide	No	
Lys	NHS-PEG-biotin	Extended amide	No	
Asp, Glu	Hz-PEG-biotin	Extended hydrazide	No	
Genetic engineering	Fusion gene elements			
	Recombinant protein, recombinant vector			No
	Recombinant vector, recombinant avidin			Flexible

A novel formulation of an exogenous plasmid DNA has been reported [75, 76], wherein either a β-galactosidase or a luciferase expression plasmid, driven by the SV40 promoter, is incorporated in the interior of neutral liposomes that are PEGylated with PEG2000. Approximately 40 of the PEG strands per liposome are tethered to the antitransferrin receptor OX26 murine mAb. It has been observed that OX26 mAb or transferrin undergoes receptor-mediated transcytosis through the BBB. This property of a peptidomimetic mAb, such as OX26 enables brain targeting of the PEGylated immunoliposomes by attachment to the TfR and subsequently internalization through the BBB.

8.2 Nanoparticles for brain delivery

Apart from engineered liposomes, nanoparticles also hold promise as delivery vehicles to transport essential drugs across the BBB that normally fail to cross this barrier. Drugs that have been successfully transported into the brain using nanoparticles include hexapeptide dalargin, dipeptide kytorphin, loperamide, tubocurarine, N-methyl-D-aspartate (NMDA) receptor antagonist MRZ 2/576 and doxorubicin [77–81]. Nanoparticles may be especially helpful for the treatment of disseminated and very aggressive brain tumors. The mechanism of nanoparticle mediated transport of drugs across the BBB is at

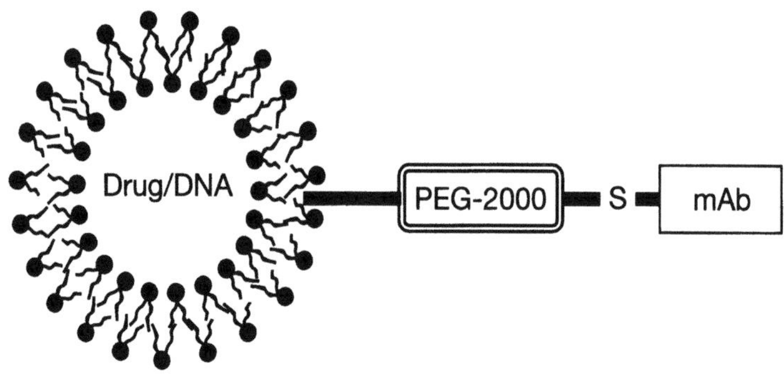

Figure 13.
PEGylated immunoliposome.

present not fully elucidated, however several possible mechanisms have been proposed. They are:

1. Higher concentration gradient of drug in the brain blood capillary due to an increase in retention of nanoparticles that would in turn enhance the transport across the endothelial cells ultimately leading to brain delivery.
2. A generalized membrane permeability enhancement of endothelial cell membrane or permeability enhancement of endothelial cell membrane caused by surfactant action.
3. An opening of tight junctions between the endothelial cells caused by nanoparticles. This would increase the delivery of free drug through the tight junction vis-à-vis higher drug payload on nanoparticles via stable binding.
4. Endocytosis of nanoparticles by endothelial cells followed by the subsequent intracellular release of drug.
5. Transcytosis of nanoparticles along with drug across the BBB.
6. Inhibition of efflux mechanism especially P-glycoprotein caused by polysorbate 80 used as a coating agent.

Nanoparticle mediated drug transport to the brain is mainly dependent on the coating of the particles with polysorbates, especially polysorbate 80. This surface active agent plays a pivotal role as an anchor for apolipoprotein E (apoE) or other substances following injection into the blood stream. The particles then probably mimic low density lipoprotein (LDL) particles and interact with the LDL receptor leading to their uptake by the endothelial cells. This may lead to release of drug in these cells and diffusion into the brain interior along with transcytosis of the particles. Other processes like tight junction modulation or P-glycoprotein (P-gp) inhibition also may contribute enhanced permeation of BBB. Furthermore, these mechanisms may run in parallel or could work in combination.

9 Conclusion

Physiological based strategies do not bypass or alter the BBB but make use of biocellular uptake mechanisms. These strategies emanated from an understanding of the anatomy and physiology based strategies including pseudonutrients, cationic antibodies and chimeric peptides. Pseudonutrients are polar micromolecular drugs that have a molecular structure mimicking a nutrient that normally permeate following carrier-mediated transport through the BBB. Cationic antibodies follow absorptive-mediated transcytosis through the BBB owing to their positive charge. The most versatile and successful approach for brain drug delivery is chimeric peptide technology, which involves a simultaneous progress in three interdependent spheres: the vector sphere, the linker sphere and the drug sphere. In addition to vector mediated transport, engineered colloidal carrier systems viz., surface modified liposomes and nanoparticles have also been investigated for brain delivery of a plethora of bioactives. In essence, these strategies offer entry to the circle of inference that the need to deliver bioactives and therapeutics to the brain using safe biomodules functional at biophysiological level could be realized.

References

1   Pardridge WM (1995) Transport of small molecules through the blood-brain barrier and methodology. *Adv Drug Delivery Rev* 15: 5–36

2   Begley DJ (1996) The blood brain barrier: principles for targeting peptides and drugs to the central nervous system. *J Pharm Pharmacol* 48: 136–146

3   Vorbrodt AW (1987) Demonstration of anionic sites on the luminal and abluminal fronts of endothelial cells with poly-L-lysine-gold complex. *J Histochem Cytochem* 35: 1261–1266

4   Vorbrodt AW (1989) Ultracytochemical characterization of anionic sites in the wall of brain capillaries. *J Neurocytol* 18: 359–368

5   Pardridge WM, Buciak JL, Kang YS, Boado RJ (1993) Protamine-mediated transport of

albumin into brain and other organs of the rat. Binding and endocytosis of protamine-albumin complex by microvascular endothelium. *J Clin Invest* 92: 2224–2229

6   Nagy Z, Peters H, Huttner I (1981) Endothelial surface charge: blood-brain barrier opening to horseradish peroxidase induced by polycation protamine sulfate. *Acta Neuropathol (Suppl)* 7: 7–9

7   Griffin DF, Giffels J (1982) Study of protein characteristics that influence entry into the cerebrospinal fluid of normal mice and mice with encephalitis. *J Clin Invest* 70: 289–295

8   Boulianne GL, Hozumi N, Shulman MJ (1984) Production of functional chimeric mouse/human antibody. *Nature* 312: 643–646

9   Neuberger MS, Williams GT, Mitchell EB, Sousley SS, Flanigan JG, Rabbits TH (1985) A hapten specific chimeric IgE antibody with human physiological effector function. *Nature* 314: 268–270

10   Lo MS, Queen C (1991) Humanized antibodies for therapy. *Nature* 315: 501–502

11   Riechmann L, Clack M, Waldmann H, Winter G (1988) Reshaping human antibodies for therapy. *Nature* 332: 323–327

12   Bruggemann M, Williams GT, Bindon CI, Clack MR, Walker MR, Jefferies R, Waldmann H, Neuberger MS (1987) Comparison of the effector functions of human immunoglobulins using a matched set of chimeric antibodies. *J Exp Med* 166, 1351–1361

13   Steplewski Z, Sun LK, Shearman CW, Ghrayeb J, Daddona P, Koprowski H (1988) Biological activity of human mouse IgG1, IgG2, IgG3 and IgG4 chimeric monoclonal antibodies with antitumor specificity. *Proc Natl Acad Sci USA* 85: 4852–4856

14   Tramentano A, Janda KD, Benkovic C, Leiner RA (1986) Catalytic antibodies. *Science* 234: 1566–1570

15   Pollock SJ, Jacobs JW, Schultz PG (1986) Selective chemical catalysis by an antibody. *Science* 234: 1570–1573

16   Coloma MJ, Morrison SL (1993) Production and expression of genetically engineered immunoglobulins. *Methods Mol Genet* 2: 209–225

17   Kostelny SA, Cole MS, Tso JY (1992) Formation of a bispecific antibody by the use of leucine zippers. *J Immunol* 148: 1547–1553

18   Hoogenboom HR, Volckert G, Raus JC (1991) Construction and expression of antibody tumor necrosis factor fusion proteins. *Mol Immunol* 28: 1027–1037

19   Butt AM, Jones HC, Abbott NJ (1989) Electrical resistance across the blood brain barrier in anaesthetized rats: A developmental study. *J Physiol (Lond)* 429: 47–62

20   Duffy KR, Pardridge WM (1989) Blood-brain barrier transcytosis of insulin in developing rabbits. *Brain Res* 420: 32–38

21   Fishman JB, Rubin JB, Handrahan JV, Connor JR, Fine RE (1987) Receptor-mediated transcytosis of transferrin across the blood-brain barrier. *J Neurosci* 18: 299–304

22   Pardridge WM (1986) Receptor mediated peptide transport through the blood-brain barrier. *Endocrine Rev* 7: 314–330

23   Duffy KR, Pardridge WM, Rosenfeld RG (1986) Human blood-brain barrier insulin-like growth factor receptor. *Metabolism* 37: 136–140

24   Bullard DE, Bourdon M, Bigner DD (1984) Comparison of various methods for delivering radiolabelled monoclonal antibody to normal rat brain. *J Neurosurg* 61: 901–911

25   Friden PM, Walus LR, Masso GF, Taylor MA, Malfroy B, Starzyk RM (1991) Anti-transferrin receptor antibody and antibody-drug conjugates cross the blood-brain barrier. *Proc Natl Acad Sci USA* 88: 4771–4775

26   Friden PM, Walus LR, Watson P, Doctrow SR, Kozarich JW, Backman C, Bergman H, Hof-

fer B, Bloom F, Granholm AC (1993) Blood-brain barrier penetration and *in vivo* activity of an NGF conjugate. *Science* 259: 373–377

27   Muckerheide A, Apple RJ, Pesce AJ, Michael JG (1987) Cationization of protein antigens. *J Immunol* 138: 833–837

28   Pardridge WM, Triguero D, Buciak JL, Yang J (1990) Evaluation of cationized rat albumin as a potential blood brain barrier drug transport vector. *J Pharmacol Exp Ther* 255: 893–899

29   Gatewood JM, Schroth GP, Schmid CW, Bradbury EM (1990) Zinc-induced secondary structure transitions in human sperm protamines. *J Biol Chem* 265: 20667–20672

30   Hardebo JE, Kahrstrom J (1985) Endothelial negative surface charge areas and blood-brain barrier function. *Acta Physiol Scand* 125: 495–499

31   Westergren I, Johansson BB (1993) Altering the blood brain barrier in the rat by intracarotid infusion of polycations: a comparison between protamine, poly-L-lysine and poly-L-arginine. *Acta Physiol Scand* 149: 99–104

32   Smith DH, Bym RA, Marsters SA, Gregory T, Groopman JE, Capon DJ (1987) Blocking of HIV-1 infectivity by a soluble, secreted form of the CD4 antigen. *Science* 238: 1704–1707

33   Hussey RE, Richardson NE, Kowalski M, Brown NR, Chang HC, Siliciano RF, Dorfman T, Walker B, Sodroski J, Reinherz EL (1988) A soluble CD4 protein selectively inhibits HIV replication and syncytium formation. *Nature* 331: 78–81

34   Pardridge WM (1991) *Peptide drug delivery to the brain*. Raven Press, New York, 1–357

35   Pardridge WM, Eisenberg J, Yang J (1987) Human blood-brain barrier transferrin receptor. *Metabolism* 36: 892–895

36   Fishman JB, Ruben JB, Handrahan JV, Connor JR, Fine RE (1987) Receptor mediated transcytosis across the blood-brain barrier. *J Neurosci Res* 18: 299–304

37   Skarlatos S, Yoshikawa T, Pardridge WM (1995) Transport of [^{125}I] transferrin through the rat blood-brain barrier *in vivo*. *Brain Res* 683: 164–171

38   Friden PM, Walus LR, Musso GF, Taylor MA, Malfroy B, Starzyk RM (1991) Anti-transferrin receptor antibody and antibody-drug conjugates cross the blood-brain barrier. *Proc Natl Acad Sci USA* 88: 4771–4775

39   Pardridge WM, Buciak JL, Friden PM (1991) Selective transport of anti-transferrin receptor antibody through the blood-brain barrier *in vivo*. *J Pharmacol Exp Ther* 259: 66–70

40   Kang YS, Pardridge WM (1994) Use of neutral avidin improves pharmacokinetics and brain delivery of biotin bound to an avidin-monoclonal antibody conjugate. *J Pharmacol Exp Ther* 269: 344–350

41   Kang YS, Pardridge WM (1994) Brain delivery of biotin bound to a conjugate of neutral avidin and cationized human albumin. *Pharm Res* 11: 1257–1264

42   Shin SU, Friden P, Moran M, Olson T, Kang YS, Pardridge WM, Morison SL (1995) Transferrin-antibody fusion proteins are effective in brain targeting. *Proc Natl Acad Sci USA* 92: 2820–2824

43   Soos MA, O'Brien RM, Brindle NPJ, Stigter JM, Okamoto AK, Whittaker J, Siddle K (1989) Monoclonal antibodies to the insulin receptor mimic metabolic effects of insulin but do not stimulate receptor autophosphorylation in transfected NIH373 fibroblasts. *Proc Natl Acad Sci USA* 86: 5217–5221

44   Soos MA, Siddle K, Baron MD, Heward JM, Luzio JP, Bellatin J, Lennox ES (1986) Monoclonal antibodies reacting with multiple epitopes on the human insulin receptor. *Biochem J* 235: 199–208

45   Pardridge WM, Kang YS, Buciak JL, Yang J (1995) Human insulin receptor monoclonal

antibody undergoes high affinity binding to human brain capillaries *in vitro* and rapid transcytosis through the blood brain barrier *in vivo* in the primates. *Pharm Res* 12: 807–816

46   Kumagai AK, Eisenberg J, Pardridge WM (1987) Absorptive-mediated endocytosis of cationized albumin and a b-endorphin-cationized albumin chimeric peptide by isolated brain capillaries. Model system of blood-brain barrier transport. *J Biol Chem* 262: 15214–15219

47   Yoshikawa T, Pardridge WM (1992) Biotin delivery to brain with a covalent conjugate of avidin and a monoclonal antibody to the transferrin receptor. *J Pharmacol Exp Ther* 263: 897–903

48   Green NM (1975) Avidin. *Adv Protein Chem* 29: 85–133

49   Green NM (1990) Avidin and streptavidin. *Methods Enzymol* 184: 51–67

50   Gitlin G, Bayer EA, Wilchek M (1990) Studies on the biotin-binding sites of avidin and streptavidin. *Biochem J* 269: 527–530

51   Morrison SL, Shin SU (1995) Genetically engineered antibodies and their application to brain delivery. *Adv Drug Deliv Rev* 15: 147–175

52   Yaksh TL, Wang JY, Go VLW, Harty GJ (1987) Cortical vasodilatation produced by vasoactive intestinal peptide (VIP) and by physiological stimuli in the cat. *J Cereb Blood Flow Metab* 7: 315–326

53   McCulloch J, Edvinsson L (1980) Cerebral circulatory and metabolic effects of vasoactive intestinal polypeptide. *Am J Physiol* 238: H449–H456

54   Bickel U, Yoshikawa T, Landaw EM, Faull KF, Pardridge WM (1993) Pharmacologic effects *in vivo* in brain by vector mediated peptide drug delivery. *Proc Natl Acad Sci USA* 90: 2618–2622

55   Itakura T, Okuno T, Nakakita K, Kamei I, Naka Y, Nakai K, Imai H, Komai N, Kimura H, Maeda T (1984) A light and electron microscopic immunohistochemical study of vasoactive intestinal polypeptide and substance p-containing nerve fibers along the cerebral blood vessels comparison with aminergic and cholinergic nerve fibers. *J Cerebral Blood Flow Metab* 4: 407–414

56   Lee TJF, Saito A (1984) Vasoactive intestinal polypeptide-like substance: the potential transmitter for cerebral vasodilation. *Science* 224: 898–901

57   Bickel U, Yamada S, Pardridge WM (1994) Synthesis and bioactivity of monobiotinylated DALDA: a μ-specific opioid peptide designed for targeted brain delivery. *J Pharmacol Exp Ther* 268: 791–796

58   Bickel U, Kang YS, Pardridge WM (1995) *In vivo* cleavage of a disulfide based chimeric opioid peptide in rat brain. *Bioconj Chem* 6: 211–218

59   Pardridge WM, Kang YS, Buciak JL (1994) Transport of human recombinant brain-derived neurotrophic factor (BDNF) through the rat blood-brain barrier *in vivo* using vector mediated peptide drug delivery. *Pharm Res* 11: 738–746

60   Wu D, Pardridge WM (1999) Neuroprotection with noninvasive neurotrophin delivery to the brain. *Proc Natl Acad Sci USA* 96: 254–259

61   Gao WY, Han FS, Storm C, Egan W, Cheng YC (1992) Phosphorothioate oligonucleotides are inhibitors of human DNA polymerases and RNase: implications for antisense technology. *Mol Pharmacol* 41: 223–229

62   Whitesell L, Geselowitz D, Chavany C, Fahmy B, Walbridge S, Alger JR, Neckers LM (1993) Stability, clearance and disposition of intraventricularly administered oligodeoxy nucleotides: implications for therapeutic application within the central nervous system. *Proc Natl Acad Sci USA* 90: 4665–4669

63 Nielsen PE, Egholm M, Berg RH, Buchardt O (1993) Peptide nucleic acids (PNAs): potential antisense and anti-gene agents. *Anti-cancer Drug Design* 8: 53–63

64 Kang YS, Boado RJ, Pardridge WM (1995) Pharmacokinetics and organ clearance of a 3'-biotinylated internally [^{32}P] labeled phosphodiester oligodeoxynucleotide coupled to a neutral avidin/monoclonal antibody conjugate. *Drug Metab Disp* 23: 55–59

65 Pardridge WM, Boado RJ, Kang YS (1995) Vector mediated delivery of a peptide nucleic acid through the blood brain barrier *in vivo*. *Proc Natl Acad Sci USA* 92: 5592–5596

66 Hanvey JC, Peffer NJ, Bisi JE, Thomson SA, Cadilla R, Josey JA, Ricca DJ, Hassman CF, Bonham MA, Au KG et al (1992) Antisense and antigene properties of peptide nucleic acids. *Science* 258: 1481–1485

67 Frenkel D, Soloman B (2002) Filamentous phage as vector-mediated antibody delivery to the brain. *Proc Natl Acad Sci USA* 99: 5675–5679

68 Wu D, Yang J, Pardridge WM (1997) Drug targeting of a peptide radiopharmaceutical through the primate blood-brain barrier *in vivo* with a monoclonal antibody to the human insulin receptor. *J Clin Invest* 100: 1804–1812

69 Kurihara N, Pardridge WM (1999) Imaging brain tumors by targeting peptide radiopharmaceuticals through the blood brain barrier. *Cancer Res* 59: 6159–6163

70 Shi N, Boado RJ, Pardridge WM (2000) Antisense imaging of gene expression in the brain *in vivo*. *Proc Natl Acad Sci USA* 97: 14709–14714

71 Shi N, Pardridge WM (2000) Noninvasive gene targeting to the brain. *Proc Natl Acad Sci USA* 97: 7567–7572

72 Gennuso R, Spigelman MK, Chinol M, Zappulla RA, Nieves J, Vallabhajosula S, Paciucci PA, Goldsmith SJ, Holland JF (1993) Effect of blood-brain barrier and blood tumor barrier modification on central nervous system liposomal uptake. *Cancer Invest* 11: 118–128

73 Micklus MJ, Greig NH, Tung J, Rapoport SI (1992) Organ distribution of liposomal formulations following intracarotid infusion in rats. *Biochim Biophys Acta* 1124: 7–12

74 Papanadjopoulos D, Allen TM, Gabizon A, Mayhew E, Matthay K, Huang SK, Lee KD, Woodle MC, Lasic DD, Redemann C et al (1991) Sterically stabilized liposomes: Improvement in pharmacokinetics and antitumor therapeutic efficacy. *Proc Natl Acad Sci USA* 88: 11460–11464

75 Li H, Sun H, Qian ZM (2002) The role of the transferrin-transferrin-receptor system in drug delivery and targeting. *Trends in Pharm Sci* 23: 206–209

76 Shi N, Pardridge WM (2000) Non-invasive gene targeting to the brain. *Proc Natl Acad Sci USA* 93: 7567–7572

77 Schroder U, Sabel BA (1996) Nanoparticles, a drug carrier system to pass the blood-brain barrier, permit central analgesic effects of i.v. dalargin injection. *Brain Res* 710: 121–124

78 Kreuter J, Alyautdin RN, Kharkevich DA, Ivanov AA (1995) Passage of peptides through the blood-brain barrier with colloidal polymer particles (nanoparticles). *Brain Res* 674: 171–174

79 Schroider U, Sommerfeld P, Ulrich S, Sebel BA (1998) Nanoparticle technology for delivery of drugs across the blood-brain barrier. *J Pharm Sci* 87: 1305–1307

80 Alyautdin RN, Petrov VE, Langer K, Berthold A, Kharkevich DA, Kreuter J (1997) Delivery of loperamide across the blood-brain barrier with polysorbate 80-coated polybutyl cyanoacrylate nanoparticles. *Pharm Res* 14: 325–328

81 Gulyaev AE, Gelperina SE, Skidan IN, Antropov AS, Kivman GY, Kreuter J (1999) Significant transport of doxorubicin into the brain with polysorbate 80-coated nanoparticles. *Pharm Res* 16: 1564–1569

Progress in Drug Research, Vol. 61
(L. Prokai and K. Prokai-Tatrai, Eds.)
©2003 Birkhäuser Verlag, Basel (Switzerland)

Peptide vectors as drug carriers

By Jamal Temsamani[1] and Jean-Michel Scherrmann[2]

[1]Synt:em
Parc Scientifique Georges Besse
30000 Nîmes, France
[2]Université René Descartes Paris 5
and INSERM U26,
Hôpital Fernand Widal
200 Rue du Faubourg Saint-Denis
75475 Paris Cedex 10
France

Jamal Temsamani

received his Ph.D. in Molecular Biology in 1989. After completing a post-doctoral fellowship at the Worcester Foundation for Biomedical Research, he joined Hybridon Inc. where he has served as Senior Research Scientist, Associate Director, and Director. In 1997, he was appointed Scientific Director of Hybridon Europe where he was responsible for supervising and coordinating European Collaborations. In 1998, he joined the company Synt:em where he is now Vice-President of Preclinical Research and Development. His research interests include drug development and drug delivery to the central nervous system. He has published more than 40 scientific articles and is co-inventor of 35 patents. He has been an invited speaker on several national and international meetings.

Jean-Michel Scherrmann

Pharm. D., Ph.D., is professor and chair, department of Clinical Pharmacy and Pharmacokinetics, Faculty of Pharmacy, The University of Paris 5. He received his Pharm. D. and Ph. D. in analytical radiochemistry from the University René Descartes at Paris. He currently leads the Neuropharmacokinetic Unit at the French Institute of Health and Medical Research (INSERM). Dr Scherrmann has made major contributions in the development of drug radioimmunoassay, drug detoxification by immunotherapy and drug redistribution concepts in pharmacokinetics. He pioneered the first clinical application of colchicine immunotherapy in acute colchicine overdoses and is now focusing his research on the role of drug transporters in neuropharmacokinetics and on drug delivery strategies to the brain. His work has resulted in two patents, 270 scientific articles, 35 book chapters and more than 250 presentations. He has served on the editorial boards of several journals in the analytical and pharmaceutical sciences. He has mentored over 28 postgraduate and 33 doctoral students. Dr. Scherrmann has been recipient of the 1992 American Academy of Clinical Toxicology Award, the 1999 French National Academy of Medicine Achievement Award and the 2002 Fellow Award of the American Association of Pharmaceutical Scientists.

Contents

Key words

Peptide-vector, drug delivery, central nervous system, blood-brain-barrier, multidrug resistance.

Glossary of abbreviations

BBB, blood-brain barrier; B-Pc, benzyl-penicillin; CNS, central nervous system; CSF, cerebrospinal fluid; β-Gal, β-galactosidase; i.v., intravenous; K_{in}, transfer coefficient; NLS, nuclear localisation signal; P-gp, P-glycoprotein; PG-1, Protegrin 1.

Table 1.
Principal cell-penetrating peptides

Cell-penetrating peptides	Sequence*	Origin	Refs.
SynB1	RGGRLSYSRRRFSTSTGR	Protegrin I	[12–14, 18]
SynB3	RRLSYSRRRF	Protegrin I	[13]
Penetratin	RQIKIWFQNRRMKWKK	Antennapedia	[2, 3, 10, 17]
Tat$_{48-60}$	GRKKRRQRRRPPQ	HIV TAT	[3, 11, 25]

* The sequence indicated in this table corresponds to the original sequence. Various analogues were described in the literature. Abbreviations: HIV, human immunodeficiency virus; TAT, transcription-activating factor.

1   Introduction

During the last decade, several peptides have been described, such as SynB vectors [1] penetratin and Tat [2, 3] that allow the intracellular delivery of polar, biologically active compounds *in vitro* and *in vivo* [2, 4]. These peptides, belonging to various families, are heterogeneous in size (10 to 18 amino acids) and sequence (Tab. 1). However, all these peptides possess multiple positive charges and some of them share common features such as important theoretical hydrophobicity and helical moment (reflecting the peptide amphipathicity), the ability to interact with lipid membrane and to adopt a significant secondary structure upon binding to lipids. The facility with which they cross the membrane into the cytoplasm even when carrying hydrophilic molecules has provided a new and powerful tool in biomedical research [3, 4]. An even more difficult task was to use these peptide vectors to deliver drugs across the blood-brain barrier (BBB). This chapter emphasizes the use of peptide vectors for brain delivery.

2   Peptide-mediated strategy

2.1   Cell-penetrating peptides

2.1.1 SynB vectors

SynB vectors are a new family of vectors derived from the antimicrobial peptide protegrin 1 (PG-1), an 18 amino acid peptide originally isolated from

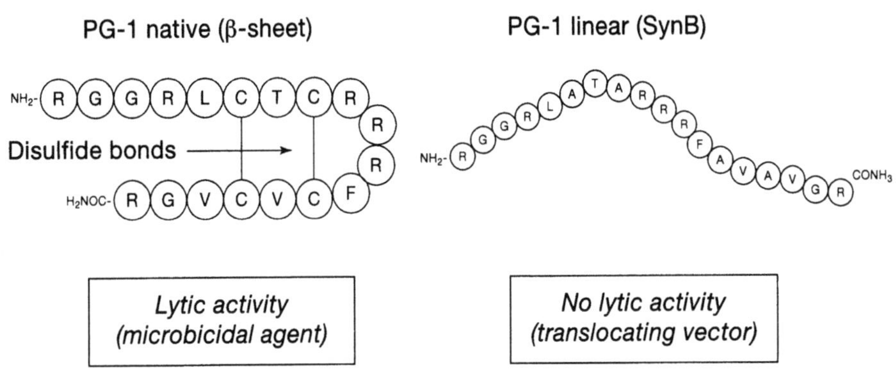

Figure 1.
Structure activity relationships of protegrins and analogues.

porcine leucocytes [5, 6]. The peptide has a β-hairpin structure in which two antiparallel strands, linked by a β-turn are stabilised by two disulphide bridges [7] (Fig. 1). As previously reported, the PG-1 peptide interacts with, and forms pores in, the lipid matrix of bacterial membranes [8, 9]. Since it has been shown that the pore formation capability of PG-1 depends on its cyclisation [9], various linear analogues of PG-1, lacking the cysteine residues, were designed (Fig. 1 and Tab. 1). These linear peptides (SynB vectors) are able to interact with the cell surface and cross the plasma membrane while their membrane-disrupting activity has been lost. Furthermore, the internalisation of these peptide vectors into cells does not appear to be dependent on a chiral receptor since the D-enantiomer form penetrates as efficiently as the parent peptide (L-form), and retro-inverso sequences exhibit identical penetrating activity. These linear protegrin analogues were the starting point for developing a new potent strategy for drug delivery into complex biological membranes such as the BBB.

2.1.2 Penetratin

It has been shown that the homeodomain of the Antennapedia protein (AntpHD), a *Drosophilia* homeoprotein, is internalised by cells in culture and is conveyed to the nucleus where it binds specifically to its DNA cognate site [2, 3, 10]. The sequence responsible for this translocation has been mapped to a region comprising the third helix of AntpHD. Furthermore, it was established that a short peptide segment, $pAntp_{43-58}$ (penetratin) corresponding to the helix itself, is able to penetrate into primary neuronal cultures [2] (Tab. 1). Additional studies showed that analogues of penetratin corresponding to its enantio- (43-58 all D) or retro-inverso form (58-43) penetrated as efficiently as the parent peptide [2, 3], suggesting that $pAntp_{43-58}$ translocates through cell membranes without binding to a stereospecific receptor and by a non-endocytotic pathway. Moreover, the uptake of peptide is not saturable and appears to be independent of cell type since it has been shown to enter into various cell lines, such as lymphocytes or endothelial cells.

2.1.3 Tat peptide

As found for the homeoproteins, the transcription factor Tat, involved in the replication cycle of the HIV was demonstrated to penetrate into cells [3, 11]. In addition, a 35 amino acid peptide corresponding to fragment 37-72 of the HIV Tat protein has been shown to promote the intracellular delivery of covalently bound proteins. Vivès et al. [11] showed that several fragments (Tat_{37-60}, Tat_{43-60}, Tat_{48-60}) derived from the most basic region of the protein penetrates into HeLa cells at 37 °C as well as 4 °C. One of the shortest peptides, Tat_{48-60} containing a nuclear localisation signal (NLS), was defined as the minimal translocating fragment, although the deletion of three non-basic residues within this sequence did not affect its cell-penetrating ability (Tab. 1) [11]. This observation underlines the role of basic residues in the translocating ability of Tat-derived peptides.

2.2   Enhancement of brain delivery

As described earlier, the BBB poses a formidable obstacle to drug therapy for the central nervous system (CNS). The fact that a peptide vector is internalized inside the cell does not guarantee that it will cross the BBB. The BBB is more complex than a simple cell layer, comprising a specialized endothelium (compared with other blood vessels) associated with pericytes and astrocyte foot processes which together elaborate a multicellular barrier. Despite this complexity, we have demonstrated that SynB peptide vectors can enhance the delivery of many different types of drugs across the BBB.

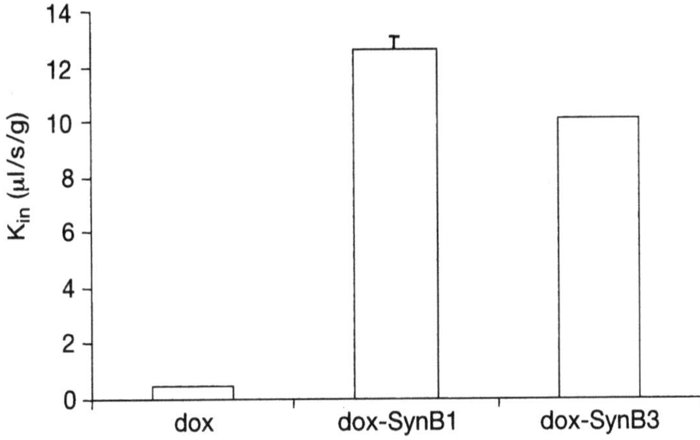

Figure 2.
Brain uptake of free and coupled doxorubicin. Transfer coefficients (K_{in}) for free and vectorised doxorubicin uptake in right hemisphere of mice brain after 60 sec perfusion with buffer. Values are mean ± SEM.

2.2.1 Small molecules

In one study, we assessed the efficacy of SynB vectors to enhance the brain uptake of the anti-cancer agent doxorubicin. Doxorubicin was conjugated to SynB vectors *via* a chemical linker (succinate) and its ability to cross the BBB was studied using *in situ* cerebral perfusion in rats and mice [12, 13]. This "vectorisation" of doxorubicin to SynB vectors significantly enhanced its brain uptake and without compromising BBB integrity (Fig. 2). The amount of vectorised doxorubicin that was delivered to the brain parenchyma (shown after applying the capillary depletion method) was about 20- to 50-fold higher than for free doxorubicin, depending on the vector used [12]. Interestingly, we also observed that SynB vectorised doxorubicin bypasses the P-glycoprotein (P-gp) which has been shown to be present in the luminal membrane of the BBB endothelial cells [14]. This 170 kDa ATP-dependent efflux pump, due to its unidirectional orientation, from brain to blood, restricts the brain entrance, or increases the brain clearance, of a broad number of therapeutic compounds, including cytotoxic drugs [15, 16]. Other experiments, carried out *in vitro* using resistant cells, have confirmed that vectorised doxorubicin bypasses the P-gp and enhances its potency in those cells [14].

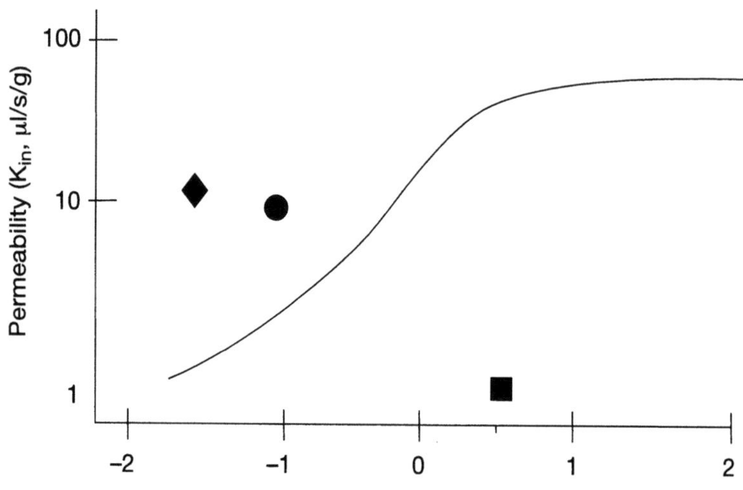

Figure 3.
Relation between log D octanol/water pH : 7.4 and blood-brain barrier permeability as determined with *in situ* brain perfusion technique for free and vectorised doxorubicin (■ free doxorubicin; ● doxorubicin-D-Penetratin; ♦ doxorubicin-Syn B1). The solid line represents the fitted relationship for compounds crossing the blood-brain barrier by only passive diffusion.

As a comparison we have also conjugated doxorubicin to D-penetratin and assessed its brain uptake by *in situ* brain perfusion [12]. A 5- to 7-fold enhancement in brain uptake was observed. The brain vascular volumes were normal at low concentrations but 2-fold larger than those observed with the SynB vectors at higher concentrations, suggesting an opening of the tight junctions [12]. Bolton et al. have shown that penetratin does not pass the BBB after an intravenous injection [17].

Figure 3 illustrates the improvement in the brain uptake of doxorubicin following its vectorization by the peptide vectors. When looking at the classical relation between the blood-brain barrier permeability *versus* the octanol-water partition, doxorubicin shows a very weak permeability despite its relative lipophilicity. The efflux from the brain endothelial cells to the blood compartment mediated by P-glycoprotein is one of the reasons for this poor brain uptake. According to the classical rules predicting that small molecular weight and lipophilic compounds can cross BBB, the coupling of doxorubicin to the peptide vectors, which increases the molecular weight from 544 Da (doxorubicin) to 3027 Da (Dox-D-Penetratin) and 2724 Da (dox-Syn B1), respectively, and markedly renders them hydrophilic should be unfavorable to an increased BBB permeability. In fact, as shown by Figure 3, both vectorised doxorubicin derivatives have an improved brain uptake which is larger than the expected values given by the fitted line for compounds crossing BBB by single passive diffusion. This new type of vectorisation of drugs using a peptide-mediated strategy tends to be localized in the same area as for hydrophobic compounds entering the brain by carrier-mediated transport. In fact, if their brain uptake mechanism at the BBB level does not imply an active transport process, it results, however, in a very efficient brain delivery.

The ability of SynB vectors to enhance the brain uptake of doxorubicin was also assessed after intravenous injection of vectorised doxorubicin into mice. The tissue and plasma distribution of doxorubicin were dramatically modified when the drug was vectorised. The brain concentrations were higher for vectorised doxorubicin compared to that of free doxorubicin [12]. Interestingly, vectorised doxorubicin shows significantly lower levels in the heart, strongly suggesting that cardiotoxicity – the main side-effect of doxorubicin – could be reduced using this strategy.

In order to assess the broader potential of this approach, we have investigated the transport of another small molecule: the antibiotic benzyl-penicillin (B-Pc) [18]. β-lactam antibiotics are often used for treatment of CNS infections but their poor penetration into the brain does not allow a sufficient efficacy. It was demonstrated that only very low concentrations of B-Pc were observed in the cerebrospinal fluid (CSF) despite high blood levels [19]. To improve its penetration across the BBB, B-Pc was coupled to SynB1 vector *via* a glycolamidic ester linker [18]. The uptake of free and vectorised [^{14}C] B-Pc to the luminal side of rat brain capillaries was measured using *in situ* brain perfusion. First, we observed by using [^{3}H] sucrose, a marker of brain vascular volume, that the integrity of BBB was not altered by the conjugate. The brain uptake of coupled B-Pc showed an average of 8-fold increase in comparison to free B-Pc (Fig. 4). This increase was quite similar for the seven explored gray areas of the rat brain. Additional experiments indicated that about 80% of the conjugated B-Pc were associated with the brain parenchyma while less that 20% were in the endothelial cells after 60 sec of perfusion. Finally, in the parenchymal brain compartment, the ratio of vectorised *versus* free B-Pc was about 7-fold [18]. Following this primary investigation, further studies are now needed to determine the fate of conjugated B-Pc within the brain parenchyma. At this level, the cleavage of B-Pc from the peptide vector in the brain appears as one of the most critical issues for observing the

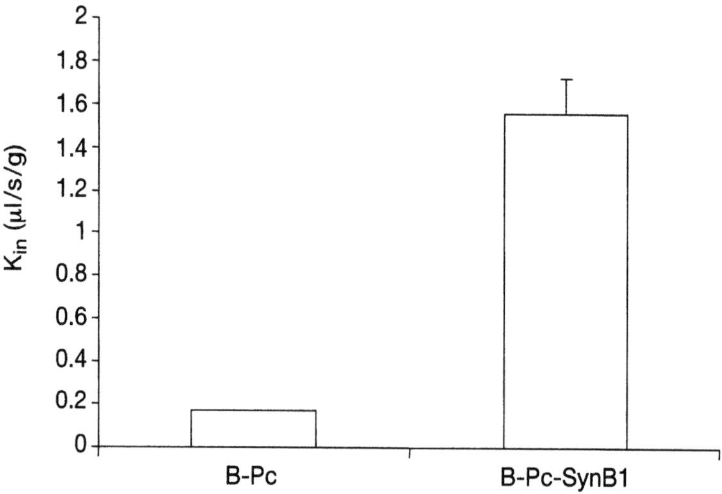

Fig. 4.
Brain uptake of free and coupled Benzylpenicillin. Transfer coefficients (K_{in}) for free and vectorised B-Pc uptake in right hemisphere of rat brain after 60 sec perfusion with buffer. Values are mean ± SEM.

therapeutic effect in an animal model. Moreover, as the brain B-Pc level depends not only on its limited entry at the BBB but also on active transport of the drug from the CSF at the choroids plexus level [20], it remains to determine if the coupled B-Pc is sensitive or not to this efflux system. Nevertheless, these data clearly demonstrated that a weak organic acid and a highly ionized compound that diffuses poorly across the BBB, such as B-Pc, can be successfully delivered to the brain parenchyma using this technology.

2.2.2 Peptides

The use of SynB vectors has also been successfully applied to brain delivery of drug-like peptide molecules. In a pharmacological application focused on pain management, the brain uptake of an enkephalin analogue dalargin was enhanced significantly after vectorisation. Dalargin is a hexapeptide analog of leu-enkephalin containing D-Ala in the second position and an additional C-terminal arginine. These modifications modulate the stability of dalargin in the blood stream and brain while at the same time modifying to some

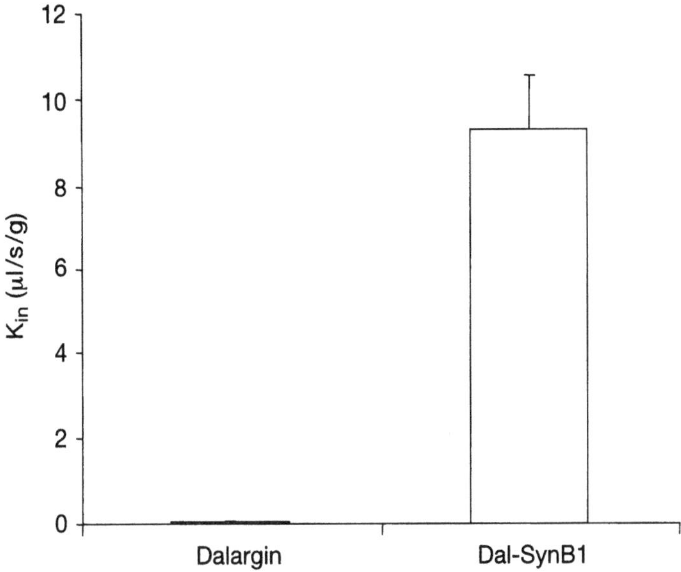

Figure 5.
Brain uptake of free and coupled dalargin. Transfer coefficients (K_{in}) for free and vectorised dalargin uptake in right hemisphere of mice brain after 60 sec perfusion with buffer. Values are mean ± SEM.

extent its receptor selectivity. While the intracerebroventricular injection of this peptide induces analgesic action, its systemic administration shows no activity in central analgesic mechanisms [21–24]. This is probably due to the poor brain uptake of dalargin. We have conjugated dalargin to SynB vectors in order to improve its brain delivery and its pharmacological effect. We have shown by *in situ* brain perfusion that vectorisation markedly enhances the brain uptake of dalargin (Fig. 5). Free or conjugated dalargin were also administered intravenously to mice and anti-nociception was determined with the Hot plate test, an assay known to be mediated by central receptors. This test measures the amount of time required for mice to react to standardized noxious stimuli. Substances which increase the reaction time are described as displaying anti-nociceptive effects, which may be interpreted as a measure of analgesia. The results show that intravenous (i.v.) administration of dalargin to mice in physiological saline exhibited no analgesic activity. In contrast, conjugation of dalargin to SynB vectors led to a considerable enhancement of analgesic activity immediately after the iv injection. For example at time 15 min post-administration, the latency time for SynB1 coupled dalargin was

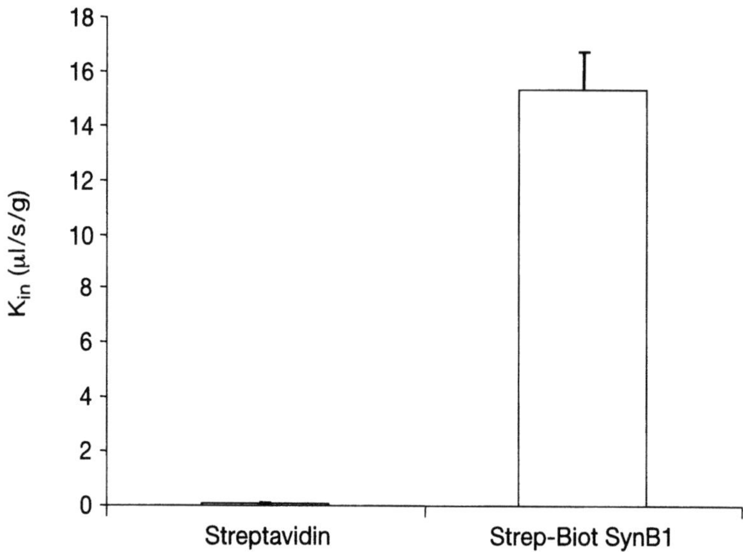

Figure 6.
Brain uptake of free and coupled streptavidin. Transfer coefficients (K_{in}) for free and vectorised streptavidin uptake in right hemisphere of mice brain after 60 sec perfusion with buffer. Values are mean ± SEM.

significantly higher than free dalargin (11.1 sec *versus* 5.8 sec) indicating an enhancement in the analgesic effect.

These results support the usefulness of peptide-mediated strategies for improving the availability and efficacy of CNS drugs.

2.2.3 Proteins

In a further application we have shown that SynB vectors are able to transport large molecules, such as the protein streptavidin (M.W. approx. 60,000 Da), across the BBB. Radiolabelled streptavidin [125I] was attached attached to a biotin moiety linked to the peptide vector. The biotinylated vector was added in large excess to allow a sufficient binding of the radiolabelled strepatvidin. *In situ* brain perfusion studies in mice showed that attachment of SynB vector to streptavidin resulted in a significant enhancement uptake into the brain (Fig. 6). As expected, free or biotin complexed streptavidin were unable to cross the BBB. These results demonstrate that this approach is able to deliver even large molecules to the brain.

Schwarze et al. [25] fused β-galactosidase (β-Gal) to Tat peptide and assessed the tissue distribution after intraperitoneal injection in mice. They observed strong β-Gal activity in all the tissues analyzed including liver, kidney, heart muscle, lung and brain. This suggested that the protein had crossed the BBB and, in addition, has passed into most other biological tissues.

2.3   Mechanism of brain uptake

We have shown by *in situ* perfusion studies, a technique allowing a first-pass exposure, that the internalisation of Dox-SynB is a saturable process [13]. The measured K_m which was in the range of 4 to 9 μM, compares well with the values observed for substrates reported to be taken up by adsorptive-mediated endocytosis. Furthermore, no difference in brain uptake was seen between doxorubicin linked to L-SynB or D-SynB vectors, indicating that a stereospecific receptor is not a requirement for its brain transport. In addition, we have reported that the passage of peptides can be inhibited in a competitive manner by polycationic molecules such as poly(L)lysine or protamine which act as endocytosis inhibitors. These observations suggest that the crossing of BBB by SynB vectors is *via* an energy dependent adsorptive-mediated endocytosis mechanism [13].

It is known that at physiological pH values the luminal surface of the brain endothelium presents an overall negative charge (due in part to significant sialylation) and thus creates an environment more selective to positively charged substances [26, 27]. The SynB peptides are positively charged. This net positive charge is likely to play a key role in the adsorptive-mediated endocytosis process wherein electrostatic interactions of the peptide vector with the surface of endothelial cells may mediate surface binding and subsequent internalization of the peptide vectors into the brain capillaries (Fig. 7).

For the transcytosis of peptides through the BBB, three steps have been proposed: 1) binding and internalization at the luminal side of endothelial cell membrane; 2) diffusion through the cytoplasm of endothelial cells; and 3) externalization at the basolateral side of endothelial cells [28]. The main components of the basal membrane are type IV collagen, fibronectin, laminins, chondroitin, and heparan sulphate from glycosaminoglycans. The most abundant component, type IV collagen, polymerizes with laminin and fibronectin proteins *via* protein binding domains such as integrin and lectin receptors [29]. These components not only provide a mechanical supporting

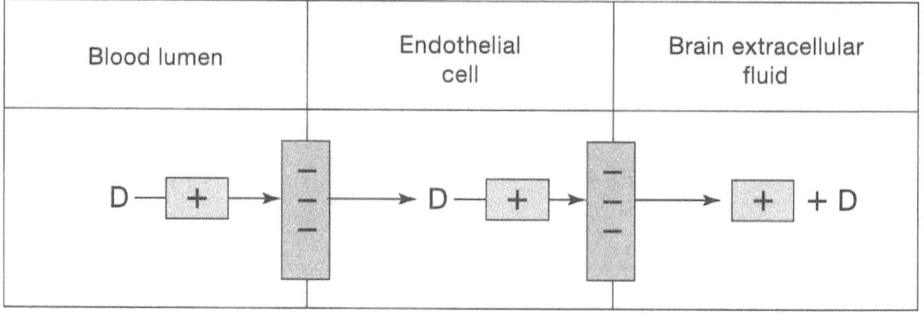

Figure 7.
Mechanism of action of adsorptive-mediated endocytosis. Positive charges of cationized peptide interact with constitutive anionic charges at the BBB (D = drug).

structure for the capillary wall, they are also important as a negatively charged barrier arising from the chondroitin and heparan sulfate residues, in addition to the anionic properties of the luminal and abluminal membranes of the endothelial cells [30]. Our results suggest that adsorptive-mediated endocytosis occurs at least at the luminal side of brain capillaries. The similarity in behavior observed for the peptide vectors studied suggests that the externalization at the abluminal side of endothelial cells may also be *via* a receptor-independent mechanism. However, since endocytosis inhibitors have only been tested at the luminal side of the endothelial cells, we cannot rule out that a different mechanism may be involved in the externalization step.

2.4 Pharmacological considerations

An additional challenge in using a peptide vector for brain delivery is the need to devise efficient ways of conjugating therapeutic molecules to the carrier (Fig. 8). Since conjugation of a drug with a transport vector may lead to a loss of biological activity, it is therefore important to develop a linker strategy that will allow the drug molecule to be cleaved from the drug transporter once it reaches its site of action. We have observed however in some cases, such as dalargin, that the conjugated drug retains a similar affinity for the its receptor as the free drug.

The coupling between the peptide vector and drug molecule may be performed by various means of chemical approaches. Direct chemical conjuga-

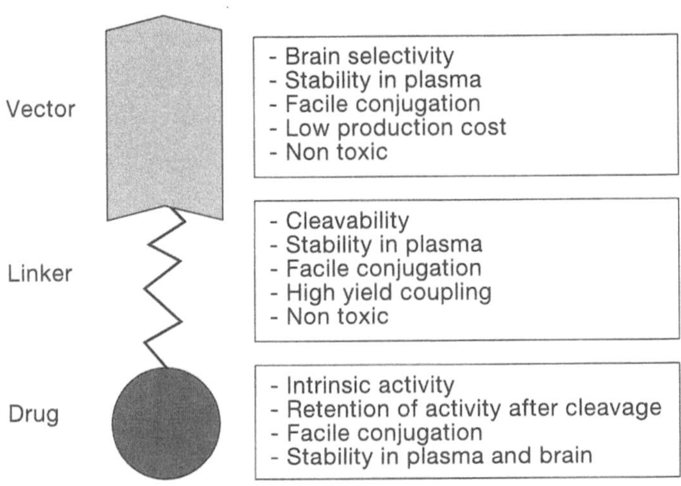

Figure 8.
Issues in the development of peptide vectors as drug carriers.

tion of vector and drug moiety has been applied for coupling of small molecules, peptides and proteins [12–14, 18, 31–33]. The disulfide-based linker may be suitable system for the vector-mediated strategy since these bonds are generally stable in plasma for several hours but are labile in brain [34].

The development of vectors for brain delivery of therapeutic molecules will not only depend on the efficiency with which these vectors cross the BBB but also on their pharmacokinetic and biodistribution profiles. If the pharmacokinetics are not optimized and the plasma bioavailability of the conjugate is low, then the brain delivery will be low. Conjugates containing the drug moiety and the vector will also need to be stable in the circulation before brain uptake occurs. Ideally, an efficient delivery system should have the dual effects of enhancing brain uptake and an increase in the systemic bioavailability of the drug in plasma.

3   Conclusion

The blood-brain barrier poses a formidable obstacle when attempting to deliver drugs to the brain. As new drugs for neurological disorders are discovered, new delivery techniques will have to be developed in concert to overcome this transport barrier. While researchers have devised many inge-nious approaches that avoid, disrupt or exploit the BBB's specialized trans-port mechanisms, many of these continue to have significant drawbacks. If neuroscientists and clinicians are to benefit from novel therapeutic advances, new delivery methods to the brain are essential. The use of peptide vectors presents a promising avenue for development. Their small size, rapid uptake, ease of drug attachment, and versatility in the range of molecules that they can deliver, provide a new approach to develop new drugs for the treatment of CNS diseases.

References

1   Temsamani J, Rousselle, C, Rees AR, Scherrmann JM (2001) Vector-Mediated drug deliv-ery to the brain. *Expert Opin Biol Ther* 1: 773–782

2   Derossi D, Chassaing G, Prochiantz A (1998) Trojan peptides: the penetratin system for intracellular delivery. *Trends Cell Biol* 8: 84–87

3   Langel U (ed) (2002) *Cell penetrating peptides. Processes and applications.* CRC Press, Boca Raton

4   Drin G, Rousselle C, Scherrmann JM, Rees AR, Temsamani J (2002) Peptide delivery to the brain *via* adsorptive-mediated endocytosis: Advances with SynB vectors. *AAPS PharmSci* 4 , article 26

5   Harwig SS, Swiderek KM, Lee TD, Lehrer RI (1995) Determination of disulphide bridges in PG-2, an antimicrobial peptide from porcine leukocytes. *J Pept Sci* 1: 207–215

6   Drin G, Temsamani J (2002) Translocation of protegrin I through phospholipid mem-branes: role of peptide folding. *Biochim Biophys Acta* 1559: 160–170

7   Aumelas A, Mangoni M, Roumestand C, Chiche L, Despaux E, Grassy G, Calas B, Cha-vanieu A (1996) Synthesis and solution structure of the antimicrobial peptide protegrin-1. *Eur J Biochem* 237: 575–583

8   Mangoni ME, Aumelas A, Charnet P, Roumestand C, Chiche L, Despaux E, Grassy G, Calas B, Chavanieu A (1996) Change in membrane permeability induced by protegrin 1: implication of disulphide bridges for pore formation. *FEBS Lett* 383: 93–98

9   Sokolov Y, Mirzabekov T, Martin DW, Lehrer RI, Kagan BL (1999) Membrane channel formation by antimicrobial protegrins. *Biochim Biophys Acta* 1420: 23–29

10   Prochiantz A (1999) Homeodomain-derived peptides. In and out of the cells. *Ann NY Acad Sci* 886: 172–179

11   Vivès E, Brodin P, Lebleu B (1997) A truncated HIV-1 Tat protein basic domain rapidly translocates through the plasma membrane and accumulates in the cell nucleus. *J Biol Chem* 272: 16010–16017

12   Rousselle C, Clair P, Lefauconnier JM, Kaczorek M, Scherrmann JM, Temsamani J (2000) New advances in the transport of doxorubicin through the blood-brain barrier by a pep-tide vector-mediated strategy. *Mol Pharmacol* 57: 679–686

13   Rousselle C, Smirnova M, Clair P, Lefauconnier JM, Chavanieu A, Calas B, Scherrmann JM, Temsamani J (2001) Enhanced delivery of doxorubicin into the brain *via* a peptide-

vector-mediated strategy: saturation kinetics and specificity. *J Pharmacol Exp Ther* 296: 124–131

14   Mazel M, Clair P, Rousselle C, Vidal P, Scherrmann JM, Mathieu D, Temsamani J (2001) Doxorubicin-peptide conjugates overcome multidrug resistance. *Anti-Cancer Drugs* 12: 107–116

15   Gottesman MM, Pastan I (1993) Biochemistry of multidrug resistance mediated by the multidrug transporter. *Annu Rev Biochem* 62: 385–427

16   Tsuji A (1998) P-glycoprotein-mediated efflux transport of anticancer drugs at the blood-brain barrier. T*her Drug Monit* 20: 588–590

17   Bolton SJ, Jones DN, Darker JG, Eggleston DS, Hunter AJ, Walsh FS (2000) Cellular uptake and spread of the cell-permeable peptide penetratin in adult rat brain. *Eur J Neurosci* 12: 2847–2855

18   Rousselle C, Clair P, Temsamani J, Scherrmann JM (2002) Improved Brain Delivery of Benzylpenicillin with a Peptide-Vector-Mediated Strategy. *J Drug Target* 10: 309–315

19   Norrby, R. (1980) Penetration of antimicrobial agents into cerebrospinal fluid: Pharmacokinetic and clinical aspects. In: JH Wood (ed): *Neurobiology of Cerebrospinal Fluid.* Plenum Press, New York, 449–463

20   Dixon RL., Owens ES, Rall, DP (1969) Evidence of active transport of benzyl-^{14}C-penicillin from cerebrospinal fluid to blood. *J Pharm Sci* 58: 1106–1109

21   Schroeder U, Schroeder H, Sabel BA (2000) Body distribution of ^{3}H-labelled dalargin bound to poly(butyl cyanoacrylate) nanoparticles after i.v. injections to mice. *Life Sci* 66: 495–502

22   Kalenikova EI, Dmitrieva OF, Korobov NV, Zhukovskii SV, Tishchenko VA (1988) Pharmacokinetics of dalargin. *Vopr Med Khim* 34: 75–83

23   Kreuter J, Alyautdin RN, Kharkevich DA, Ivanov AA. (1995) Passage of peptides through the blood-brain barrier with colloidal polymer particles (nanoparticles). *Brain Res* 674: 171–174

24   Schroeder U, Sommerfeld P, Ulrich S, Sabel BA (1998) Nanoparticle technology for delivery of drugs across the blood-brain barrier. *J Pharm Sci* 87: 1305–1307

25   Schwarze SR, Ho A, VoceroAkbani A, Dowdy SF (1999) *In vivo* protein transduction of a biologically active protein in the mouse. *Science* 285: 1569–1572

26   Hardebo JE, Kahrstrom J (1985) Endothelial negative surface charge areas, and blood-brain barrier function. *Acta Physiol Scand* 125: 495–499

27   Vordbrodt AW (1988) Ultrastructural cytochemistry of blood-brain barrier endothelia. *Prog Histochem Cytochem* 18: 1–99

28   Bar RS, DeRose A, Sandra A, Peacock ML, Owen WG (1983) Insulin binding to microvascular endothelium of intact heart: A kinetic and morphometric analysis. *Am J Physiol* 244: 477–482

29   Virgintino D, Monaghan P, Robertson D, Errede M, Bertossi M, Ambrosi G, Roncali L (1997) An immunohistochemical and morphometric study on astrocytes and microvasculature in the human cerebral cortex. *Histochem J* 29: 655–660

30   Bertossi M, Virgintino D, Maiorano E, Occhiogrosso M, Roncali L (1997) Ultrastructural and morphometric investigation of human brain capillaries in normal and peritumoral tissues. *Ultrastruct Pathol* 21: 41–49

31   Walus LR, Pardridge WM, Starzyk RM, Friden PM (1996) Enhanced uptake of rsCD4 across the rodent and primate blood-brain barrier after conjugation to anti-transferrin receptor antibodies. *J Pharmacol Exp Ther* 277: 1067–1075

32   Friden PM, Walus LR, Musso GF, Taylor MA, Malfroy B, Starzyk RM (1991) Anti-transferrin receptor antibody and antibody-drug conjugates cross the blood-brain barrier. *Proc Natl Acad Sci USA* 88: 4771–4775

33   Friden PM, Walus LR, Watson P, Doctrow SR, Kozarich JW, Backman C, Bergman H, Hoffer B, Bloom F, Granholm AC (1993) Blood-brain barrier penetration and *in vivo* activity of an NGF conjugate. *Science* 259: 373–377

34   Letvin NL, Chalifoux LV, Reimann KA, Ritz J, Schlossman SF, Lambert JM (1986) *In vivo* administration of lymphocyte-specific monoclonal antibodies in nonhuman primates. *In vivo* stability of disulfide-linked immunotoxin conjugates. *J Clin Invest* 77: 977–984

Index

MIX
Papier aus verantwortungsvollen Quellen
Paper from responsible sources
FSC® C105338

If you have any concerns about our products,
you can contact us on
ProductSafety@springernature.com

In case Publisher is established outside the EU,
the EU authorized representative is:
**Springer Nature Customer Service Center GmbH
Europaplatz 3, 69115 Heidelberg, Germany**

Printed by Libri Plureos GmbH
in Hamburg, Germany